CLEANING UP

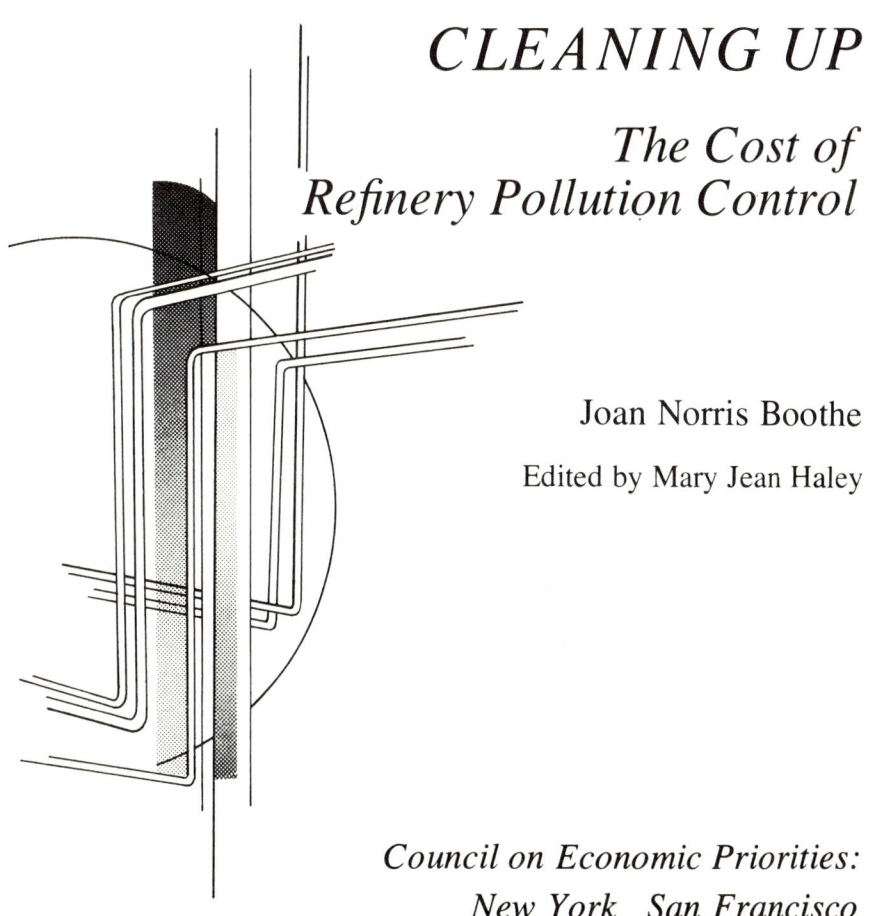

CLEANING UP

The Cost of Refinery Pollution Control

Joan Norris Boothe

Edited by Mary Jean Haley

Council on Economic Priorities:
New York   San Francisco

Joan Norris Boothe earned her BA in Economics at Wellsley, went on to do graduate work in Economics at Columbia University, and recently completed her MBA in Finance at the University of California, Berkeley. While assistant to the economist at Goldman, Sachs & Co. she wrote "Inflation: A Primer for Investors."

The author would like to give special thanks to Barry Boothe, Frederick Balderston, Mary Jean Haley, Gregg Kerlin, Daniel Rabovsky and Kathy O'Brian for the aid, advice and comfort they offered during the preparation of this study.

Illustration and production by Jaime Robles.

International Standard Book Number: 0-87871-002-7
Library of Congress Catalog Card Number: 75-10535
©1975 by the Council on Economic Priorities.
All rights reserved.

Council on Economic Priorities,
84 Fifth Ave., New York, 10011.
Published in 1975 by the Council on Economic Priorities, Inc.

Printed in the United States of America.

TABLE OF CONTENTS

SUMMARY OF FINDINGS	1
CHAPTER 1: Cleaning Up	5
CHAPTER 2: The Pocketbook Issue	17
CHAPTER 3: What *Are* Pollution Control Costs?	33
CHAPTER 4: Calculating Costs	41
CHAPTER 5: The Big Eight Refiners	51
REFINERY PROFILES:	
Atlantic Richfield Corporation	67
Exxon Corporation	71
Gulf Oil Corporation	75
Mobil Oil Corporation	79
Shell Oil Corporation	82
Standard Oil of California	86
Standard Oil Company (Indiana)	90
Texaco, Inc.	93
APPENDIX	97
GLOSSARY	110

SUMMARY OF FINDINGS

Eight major petroleum companies dominate the oil refining business in the United States. The big eight—Exxon, Shell, Standard Oil (Indiana), Texaco, Standard Oil of California, Mobil, Gulf, and Atlantic Richfield—are vertically integrated firms with interests in every aspect of the oil business. They search for oil, pump and transport it, refine it, transport it again, and market the final product. The refinery is the key link in this chain from wellhead to customer, and the big eight refiners account for over 55% of the industry's total capacity.

The refining industry as a whole shipped around $30 billion worth of products in 1973, an amount equal to almost 2.3% of the Gross National Product. Vital as refining is to the nation's economy and energy production, it can also be a major source of pollution. Oil refining is a very dirty business when it is uncontrolled.

In this study we examine the costs of controlling refinery pollution, both for the entire industry and for the eight major companies. The big eight operated a total of 61 finished fuels refineries in 1974. These refineries had a combined capacity of nearly eight million barrels a day (bbl/d) and an average size of 130,000 bbl/d. The greatest difference between the big eight and the rest of the industry is the average big eight refinery is nearly four times as large as the average refinery owned by other refiners. They also tend to be more complex than the average and more concentrated in heavily industrialized and populated regions. All of this makes pollution control by the big eight industry leaders especially important, and their practices may be expected to set the standard for the rest of the industry.

This study is a companion to a larger CEP study, *Cracking Down: Oil Refining and Pollution Control,* which examines the pollution control records of the 61 refineries operated by the eight major refiners. We used the results of the companion study to determine which of the big eight firms faces the largest pollution control costs in the future, and which the smallest. CEP found that the total industry costs for raising pollution control standards to comply with government regulations are huge, but not in proportion to the industry itself. We find that the answer to the question, "Can the economy afford the extra dollars and energy required to meet petroleum refining pollution control goals?" is YES. The industry refiners can control their pollution for well under a half cent a gallon of product with little or no net additional expenditure of energy. A more specific summary of the conclusions of the study follows:

Pollution Control Investment: CEP estimates that the industry will have to invest $3,350 million for pollution control equipment between 1974 and 1983. Approximately $2,000 million of this is to bring existing refineries up to standard. The remainder is needed to make refinery expansion that is assumed will be built between now and 1983 environmentally acceptable.

Increased Product Costs: It will cost from $5,900 million to $7,500 million more to produce petroleum products from 1974 through 1983 if refineries meet government pollution control standards than it would if pollution control quality were not improved above 1973 levels. These costs amount to an additional 11-14¢ for each barrel of product, or 0.3¢ a gallon. In the long run, refining costs will rise by 12-16¢ per barrel above 1973 costs.

Price Increases Versus Profit Reductions: Even if 100% of the increased costs were passed along to the customer, petroleum product prices would be raised by less than a TOTAL of 1% over the ten year period. If there were no price increases, oil company profits for this period could fall by 2-3%. It is most likely that the increased costs will be shared between the refiner and the customer with the majority of the burden falling on the customer.

Costs and The Companies: The present level of pollution control varies widely among the eight firms, and that means that each firm will have to spend a greater or lesser amount of money in order to complete the job by 1983. CEP estimates that post-1974 pollution control will cost Gulf, the company with the most left to spend per barrel of capacity, 40% more per barrel than Atlantic Richfield, the firm with the smallest future bill. Probable 1974-83 profit reductions range from a low of 0.1% for Exxon to a high of 1+% for Gulf. The summary table shows how the eight companies compare to one another with respect to cost increases above 1973 levels and profit reductions for the 1974-83 period.

1974-1983 COST INCREASE PER BARREL

Least Cost Increase (9-11¢):	Atlantic Richfield *(Arco)*
.	Shell
.	Exxon
Industry Average (11-14¢)	Standard Oil (Indiana) *(Amoco)*
.	Standard Oil of California
.	Texaco
.	Mobil
Most Cost Increase (14-17¢)	Gulf

1974-1983 PROFIT REDUCTION

Least Profit Reduction (.1%):	Exxon
.	Atlantic Richfield (Arco)
.	Shell
.	Texaco
.	Standard Oil of California
.	Mobil
.	Standard Oil (Indiana) *(Amoco)*
Most Profit Reduction (1.0%)	Gulf

CEP arrived at some general conclusions which are less susceptible to statistical expression, but which are nonetheless valid:

Energy and Refining: Energy conservation and efficient energy use within a refinery are not necessarily at odds with pollution control. Reducing the amount of fuel gas and oil a refinery uses to process crude not only reduces energy use, it is a very effective way to reduce the air pollution that fuel burning creates. Several other forms of pollution control either provide energy to the refinery, reduce net energy use, or reduce the amount of energy the refinery must buy from outside sources.

Returns to Control: Petroleum refining pollution control brings returns to the public and the refiner. The costs of pollution damage to health and property have never been completely quantified, so the returns from reducing that damage are equally difficult to quantify; the benefits of a cleaner, healther environment to live in are, however, real. Refiners themselves also benefit from pollution control. Air pollution control in particular can generate revenue in the form of salable recovered materials, or it can reduce refining costs (see Chapter 3). This is less true for water pollution abatement at the present time, although limiting oil and grease pollution means that more crude oil is saved to be refined. Research may soon turn today's pollutant into tomorrow's product.

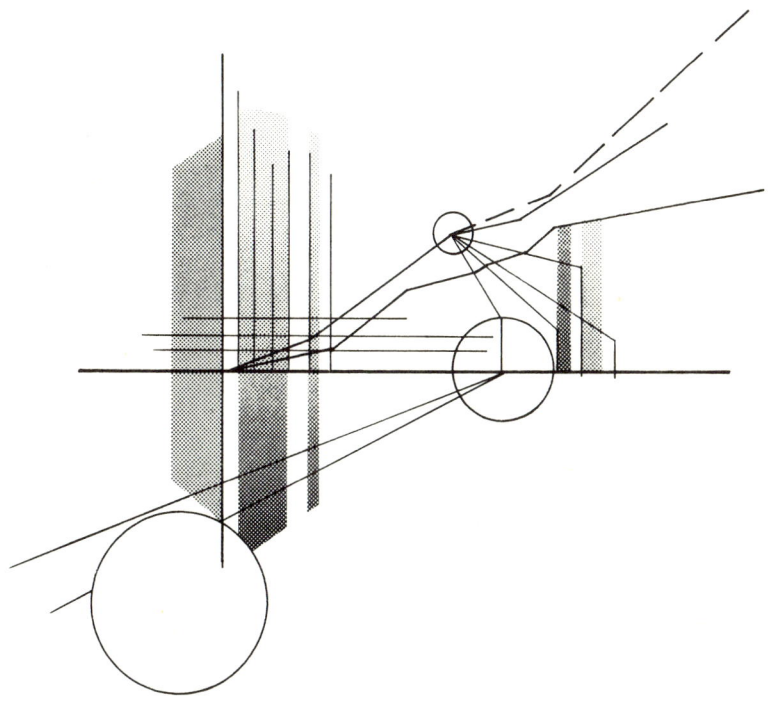

1: Cleaning Up

Environmental standards and pollution control goals are under attack. Petroleum refining pollution control regulations are a case in point. Cleaning up this industry, some now charge, is too expensive in an energy-short, inflation ridden era. Refining is vital to the nation's energy production, and it can be a major source of pollution. This study attempts to provide a perspective on the true costs of environmentally acceptable refining, both in dollar and in energy terms. We find that the answer to the question, "Can the economy afford the extra energy and dollars required to meet pollution control goals?" is YES for the petroleum refining industry. The refiners can control their pollution for well under half a cent per gallon of product.

Such a question would have been unlikely to arise in the late 1960's and early 1970's when concern over environmental degradation was widespread. Backed by enthusiastic public support, federal, state and local governments passed laws requiring business to clean up production processes and stop fouling the air and water. Nationwide, the Clean Air Act of 1970 and the Federal Water

Quality Act Amendments of 1972 set the clean-up deadlines that polluters will have to meet in the next decade.

In the mid 1970's, however, the energy crisis and inflation have become major issues, and the public recognizes that pollution control is not free. Some think there must be an automatic conflict between a clean environment and the newly urgent goals of adequate energy and stable prices. But those who simplistically declare that industry and the economy cannot afford pollution control are just as much in error as those who earlier ignored the costs and trade-offs that pollution control involves.

Some level of pollution control *is* affordable. Government standards attempt to define that level by setting control goals in terms of the technology that is available. (This is especially true for water quality standards.) What is available, though, is not necessarily affordable. Only a cost analysis done in terms of the polluter's costs of production, profits, and product prices can begin to determine affordability. CEP has attempted to do this. To be complete, such an analysis would have to include the recognition that uncontrolled pollution creates real costs too. CEP has not been able to include the costs of pollution damage in this analysis. These costs are difficult to measure, but that does not mean that they are nonexistent or insignificant.

The goals government has set for controlling petroleum refining pollution are not impossibly expensive for the industry or the economy. CEP finds that, while it will cost a great deal for the industry to comply with government regulations—especially those for water quality—it will not be so expensive that the good of the economy requires a halt in progress toward meeting those standards. Not even a slowdown is in order. At most, the expenditures involved would add only 2-3% to the present prices of petroleum products.* Furthermore, petroleum refining pollution control is not necessarily at odds with energy conservation. Indeed, we find that in many cases, just the opposite is true. Good pollution control and energy conservation frequently go hand in hand in this industry.

THE SCOPE OF THE STUDY

This study is concerned only with the costs of refinery pollution control, the costs to the refining industry as a whole, and the effect that control costs will have on the finances and profits of the eight major US refiners—Exxon, Shell, Texaco, Mobil, Standard Oil (Indiana) (also known as Amoco), Gulf, Standard Oil of California (Socal), and Atlantic Richfield (Arco).

We have not dealt with the pollution problems caused by the use of petroleum products, the question of pipelines, or any other of the topics commonly associated with the petroleum industry. The study was undertaken as a compan-

---

*All dollars and calculations in this report are in terms of constant 1974 dollars. The possible effects of inflation or other influences on prices are not considered.

ion to a much larger work, *Cracking Down: Oil Refining and Pollution Control* (1), which investigates the actual pollution control performance of the eight major refiners. Table 1 lists the eight companies included in both studies and their US refining capacity as of January 1, 1974.

TABLE 1

The Big Eight Refiners

Company	Total US Refining Capacity (bbl/d)*	% of Total US Refining Capacity
Exxon	1,252,000	8.8%
Shell	1,127,000	7.9%
Standard Oil (Indiana) *(Amoco)*	1,043,000	7.3%
Texaco	1,037,000	7.3%
Standard Oil of California *(Socal)*	952,000	6.7%
Mobil	887,000	6.2%
Gulf	860,000	6.1%
Atlantic Richfield *(ARCO)*	785,000	5.5%
Rest of Industry	6,257,000	44.0%
TOTAL	14,216,000	100.0%

*One barrel equals 42 gallons.
Note: Figures as of January 1, 1974.

Taken together, these eight firms account for over 55% of US refining capacity.* They operate a total of 61 finished fuels refineries in 22 states. The figures in Table 1 exclude some very small refineries which produce only asphalt. When full product range refineries only are included, these eight, vertically integrated petroleum companies control 60% of the country's finished fuels production. The refineries range in size from the 4,000 barrels a day (bbl/d) Kenai, Alaska refinery run by Socal to the 450,000 bbl/d giant which Exxon

*Such concentration is the rule in the US economy. The four largest steel companies hold 60-70% of industry capacity. The four largest auto makers produce nearly 80% of the cars sold. Seventy-one per cent of metal cans come from four companies, and 81% of cigarettes from another four firms (9).

operates at Baton Rouge, Louisiana. The companion study includes both pollution control performance at individual refineries and a company by company comparison. The summary of findings and conclusions from that study is reproduced as Table 2.

TABLE 2

Summary of Conclusions from
Cracking Down: Oil Refining and Pollution Control

CEP examined the pollution control performance of the eight integrated petroleum companies which dominate oil refining in the United States. CEP used emissions and discharge data from 1972, 1973, and 1974 to study the pollution control performance of each of these companies 61 refineries. An examination of the control performance for four air pollutants—sulfur oxides (SOx), carbon monoxide (CO), particulates, and hydrocarbons—and five water pollutants——biochemical oxygen demand (BOD), chemical oxygen demand (COD), oil and grease, phenols, and ammonia—formed the basis for the following conclusions:

1. Atlantic Richfield (Arco) has the best overall pollution control record of the eight. Texaco and Gulf displayed the poorest overall control performances. Arco does the best air pollution control job and a reasonable water quality control job. Shell has the best water control record, but its mediocre air pollution control performance gives the company a lower overall ranking than Arco. Texaco was the worst air polluter, Gulf the worst water polluter. The relative rankings for all of the companies follow.

COMPANY RANKING: RELATIVE OVERALL PERFORMANCE

	RELATIVE WATER PERFORMANCE	*RELATIVE AIR PERFORMANCE*
Best:	Atlantic Richfield	Shell
.	Standard Oil, Calif.	Exxon
.	Mobil	Atlantic Richfield
.	Gulf	Standard Oil (Indiana)
.	Shell	Texaco
.	Exxon	Standard Oil, Calif.
.	Standard Oil (Indiana)	Mobil
Worst:	Texaco	Gulf

TABLE 2, continued.

The range from best to worst is wide enough for CEP to describe Arco's air performance as "good" and Texaco's as "very poor." Similarly, Shell's top rated water control is also "good," and Gulf's worst water pollution control record is "poor."

2. CEP believes that the attitudes and abilities of the companies themselves are the most important source of the performance differences. Nonetheless, CEP found that good company performance in air pollution control did not guarantee good performance in water pollution control, and vice versa. In addition, it was not unusual for a single company to operate a very clean refinery in one locale and a very poorly controlled refinery in another.

3. The discrepancy from refinery to refinery within a given company can be explained in part by differences in regulation and enforcement from state to state. CEP found that all companies tend to perform well under strong pollution control agencies. When regulations and enforcement are weak, all refineries tend to perform poorly. Illinois refineries had the poorest air pollution control records in general; Los Angeles plants, on the other hand, usually controlled air pollution well. The locales switch roles for water quality control. Illinois and Washington get top marks for good water pollution control. Los Angeles and New Jersey refineries are among the worst water polluters.

4. A refinery's design, the products it makes, and the composition of the crude oil it processes all affect the difficulty of the pollution control job it faces. Even so, CEP found that there is no correlation between the sulfur content of the crude oil processed and a company's SOx control record. At the same time, size and complexity cannot be used to explain a refinery's water pollution control record. CEP found that those refineries which performed best in water pollution control did so because they had installed adequate treatment facilities and not because they had the easiest water pollution control task.

A LINK IN THE CHAIN

Petroleum refining is only part of a larger whole, the giant petroleum industry. Before it reaches the customer, crude oil must be discovered, pumped from the ground, transported, refined, transported again, and finally marketed. The refinery is a link in the middle of this chain of events. Approximately 85% of the United States' domestic refining capacity is owned by integrated petroleum companies (2), firms which participate in all aspects of the petroleum business from the original well drilling to the filling of the motorist's gasoline tank.

Many of these companies are industrial giants. Table 3 indicates how the eight companies which control the majority of US refining compare with other US industrial corporations. Clearly, we are considering firms which control a major portion of the US economy.

TABLE 3

Oil Companies and Their Peers

The Twenty Largest U.S. Industrial Corporations in 1973, Ranked by Assets

(All figures are in millions of dollars.)

Company	Assets	Rank	Sales	Rank	Profits	Rank
Exxon	25,079	1	25,724	2	2,443	1
General Motors	20,297	2	35,798	1	2,398	2
Texaco	13,595	3	9,802	10	1,292	4
Ford	12,954	4	23,015	3	906	5
IBM	12,289	5	10,993	7	1,575	3
Gen. Tel. & Elec.	10,749	6	5,105	18	352	14
Mobil	10,690	7	11,390	6	849	6
IT&T	10,133	8	10,183	8	528	12
Gulf	10,074	9	8,417	11	800	8
Standard Oil California	9,082	10	7,762	12	844	7
General Electric	8,324	11	11,575	5	585	11
Standard Oil (Indiana)	7,018	12	5,416	16	511	13
U.S. Steel	6,918	13	6,952	14	326	17
Chrysler	6,105	14	11,774	4	255	26
Tenneco	5,427	15	3,910	20	230	32
Shell	5,381	16	4,884	19	333	16
Atlantic Richfield	5,109	17	3,983	26	270	24
Du Pont	4,832	18	5,275	16	586	10
Western Electric	4,828	19	7,037	12	315	18
Westinghouse Electric	4,407	20	5,702	14	162	44

Refining itself comprises only 15% of the petroleum industry's total domestic assets (4), but that is 15% of a gigantic total. All by itself, refining is big business. Almost 250 refineries were operating in the United States in 1974. Their average capacity was about 58,000 bbl/day. At 1974 construction prices, it would cost more than $100,000,000 to build such an average-sized refinery (5). The behemoths of the industry range up to 450,000 bbl/day and probably would cost nearly $1 billion to build from scratch.

Accurate financial figures specifically related to petroleum refining are scarce because, as the Federal Energy Administration and the Department of Defense have learned, most integrated petroleum companies are close-mouthed about the details of their operations. One must rely on the rough estimates that can be and have been made. The 1973 value of industry shipments was on the order of $30 billion, almost 2.3% of the Gross National Product (6). The industry produced these products using plant and equiment with a replacement value in the realm of $16-17 billion (7), about 1% of the value of US business capital in 1973.* That year, 150,000 people, or only 0.15% of the labor force (7,6), worked in these plants. Figures for refining profits alone are generally elusive on an industry-wide basis because they usually arise as part of integrated company operations. Petroleum company books often report zero or very small profits from refining operations.

In fact, this is more an accounting device than a reflection of the importance of refining to total company operations. Refining is a vital part of the process by which petroleum companies profit from crude oil production. Unrefined crude oil is of little use to modern technology. It is the refinery which turns the raw material into a product which the company can sell. John D. Rockefeller's original Standard Oil Company was built on this very premise—the most important link in the chain is the refinery. He ignored crude oil production almost entirely in the early history of his company and only later expanded its operations to pipelines and production.

It is therefore reasonable to argue that refineries should be assigned an important share of petroleum industry profits. If we assume that refiners earn about the same return on investment as the petroleum companies in general, 1973 profits assignable to refining operations are in the realm of $2 billion (8). Further support for such a figure comes from comparing the return on sales earned by the few independent refiners to the industry's total shipments. Again, the estimate comes out in the $2 billion realm (9).

The conclusions to be reached by examining the level of refining industry profits, and the possible reduction that pollution control costs could cause, are of

---

*Throughout this study we shall be using the terms "capital" and "capital equipment" in the economists' sense of one of the factors of production: land, labor, and capital. From a production point of view, these terms mean the actual plant and equipment used in the production, or control, process. Financially, they denote the monies tied up in the physical plant and equipment. The two meanings are really both sides of the same coin and both generally apply. Other economic and technical terms are defined in the Glossary.

interest chiefly in determining the impact pollution control costs will have on independent refiners and in deciding how strong an incentive to overseas expansion domestic pollution control regulations actually provide to multinational firms. Although vertically integrated firms own the vast majority of US refining capacity, refining is not their major business. Of the eight major refiners, Shell and Arco have the largest stake in domestic refining (10), but this investment still makes up under 20% of total company fixed investment. Exxon, by far the largest petroleum company, as well as the largest refiner, has committed under 4% of its total corporate investment to domestic refining. For such companies, any effect that domestic refinery pollution control has on total company profits will be more a function of the proportion of investment in domestic refining than of the impact added costs will have on refining profits as such.

Increased US costs or reduced refining profits could lead some multinational petroleum firms to build more refining capacity abroad. This could be of real concern, since US imports of refined products (in addition to crude oil) have grown quite rapidly to close the gap between lagging refinery expansion and accelerating product demand. Since the 1940's, domestic refining capacity has expanded at a fairly steady 3.5% a year (4). Growth slowed in the late 1960's and early 70's, and many current projections indicate that future growth will continue to be slow (11). Up to the present, the slowdown in capacity expansion has not been caused by a sluggish demand for refined products. Rather, many companies have concentrated their expansion efforts outside the United States and then imported their products to this country. While there are many reasons for the slowed pace of domestic refinery expansion, the industry and some industry observers have placed a significant portion of the blame on environmental restrictions and pollution control costs. CEP concludes, however, that rational refinery location decisions should not be seriously affected by future costs of domestic pollution control per se.*

Analysts who consider pollution control costs to be a deterrent to locating new plants in the US generally assume that overseas capacity will not need similar controls. This may have been true in the past, but other countries are becoming more environmentally conscious. Japan, for example, is implementing stringent pollution control regulations, and many of the European Common Market countries are following suit (12). Thus, pollution control costs in the United States may well cease to be even a minor hindrance to domestic location in the future. And there are great advantages to being located here, particularly closeness to markets. In general it is more difficult and costly to transport finished products than crude oil.

*Environmentally related site restrictions may be another matter. Shell, Exxon, and several independent refiners have had problems with site approvals. The widely publicized case of Onassis' attempt to build a refinery in New Hampshire is a case in point. The profits that are foregone when a refinery cannot be built are real pollution control costs if the money cannot be invested equally profitably elsewhere. It is more likely, however, that large refiners that already have refineries can divert at least some expansion to enlarging an existing plant. Historically, 80% of industry expansion has taken place at existing refineries, and only 20% at new sites.

A TECHNICAL DIGRESSION

An adequate understanding of a refiner's environmental costs requires some feel for what happens in a refinery, where the pollutants arise in the processing, just what they are, and how they can be and are being controlled. Controlling pollution from any production process is seldom a simple matter of trapping various types of dirt before they get away from the plant. The polluter may have to alter processes, redesign control equipment, or undertake a major plant overhaul. The types of costs that arise are directly related to the control methods used. In addition, many aspects of pollution control bring returns to the manufacturer. As we shall see in Chapter 3, the true net costs of control are often much smaller than they appear at first. The next few pages digress briefly from the main focus of this study to provide a minimum technical description of the refining process, sources of pollutants, and types of controls.

When crude oil enters a refinery it is a heterogenous mixture of hydrocarbons and various impurities. No two crude oils are exactly alike. Crude is a brownish-green to black liquid made up of 83-87% carbon, 11-14% hydrogen, some sulfur, oxygen, nitrogen, and trace metal compounds (13). Widely differing types of crude require different refining techniques and yield different product mixes.

The chemistry of petroleum refining is highly complex. To produce refined products, the refinery separates the hydrocarbons in the crude into select groupings or fractions, converts some hydrocarbons into more salable types, and removes impurities to improve product quality and performance. This is the basic work of the refinery.

The crude is first mixed with water in a desalter to remove the soluble impurities. The second step is usually some form of distillation. The crude is heated and pumped into a distillation tower where the lightest hydrocarbons rise to the top of the tower, the medium sized ones distribute themselves throughout the tower according to their weight, and the heaviest settle to the bottom. Once the hydrocarbons are separated into fractions according to their weight, they can be drawn off for further processing. Distillation takes place several times during refining.

Each refinery has a different processing scheme geared to the crude it refines and the products it makes. However, the fluid catalytic cracker (FCC) is the heart of most modern refineries because it is central to the production of gasoline. Cracking breaks large hydrocarbon molecules into more commercially desirable sizes by using heat, pressure, and a catalyst. An FCC is divided into a reactor and a regenerator. In the reactor, a catalytically induced reaction cracks the molecules. The reaction covers the catalyst particles with coke. The coke-covered catalyst flows into the regenerator where the coke is burnt off. This both regenerates, or cleans, the catalyst, and heats it to the proper temperature before it flows back to the reactor.

Once the hydrocarbons in the crude have been separated into fractions and the unusable molecules have been cracked or otherwise reformed, the products receive a final treatment. There are many final treatment processes. Some use solvents to remove impurities from the product. Others use filters. There are, of course, product additives. The infamous tetra-ethyl lead that is added to some gasolines to raise their octane rating is one example.

Each refining process is potentially polluting. From a pollution control perspective, the significant aspect of a crude oil is the amount of sulfur, nitrogen, and trace metals it contains. Since these are considered impurities in most refined products, they often leave the refinery as pollutants.

Most refinery pollutants originate in the crude, and simple good housekeeping practices can prevent much pollution before it starts. Careful storage and handling can minimize spills and leaks which represent a loss of materials as well as pollution. Further loss through evaporation can be controlled by vapor recovery systems and by floating roof storage tanks which prevent the formation of a vapor layer.

The most important air pollutants refining produces are sulfur oxides (SOx) and hydrocarbon gases. Both can be health hazards and they represent a loss of materials if they go up the stack to become air pollutants. SOx is produced when sulfur-containing fuels are burned. SOx is difficult to control directly, but its formation can be prevented if fuels are desulfurized before they are burned. The sulfur compounds taken from fuels can be sent to a sulfur recovery plant and the recovered sulfur can be sold. Hydrocarbon gases, a component of photochemical smog, are produced by several refining processes. These process gases can be recovered and used as refinery fuel or sold for use elsewhere.

Refining also produces large amounts of nitrogen oxides (NOx), carbon monoxide (CO), and particulates. NOx is another product of fuel combustion and a major component of photochemical smog. It is very difficult to control with present technology. CO pollution from refineries has been largely eliminated through the use of CO boilers which burn the CO to produce steam that is used to power other refinery processes. CO boilers can also burn process gases from the FCC. Much of the particulate pollution a refinery produces is made up of catalyst particles. Refining catalyst is usually very expensive, and the larger particles are well worth recovering. Particulate is captured by simple mechanical devices called cyclones, which operate on the principle of centrifugal force, and by large electrostatic precipitators.

Catalytic cracking produces more air pollution than any other single refinery process. FCC exhaust contains large amounts of SOx, particulate matter, hydrocarbon gases, NOx and CO. If the feed to the cracker is desulfurized by hydroprocessing, much SOx pollution can be prevented. Hydroprocessing replaces sulfur and other impurities that are attached to hydrocarbon molecules with hydrogen. The displaced impurities often take the form of sulfurous gases which should be sent to a sulfur recovery plant.

Like the air pollutants, most refinery water pollutants are components of the crude itself. Many of them can be directly toxic to aquatic life, while others exert what is called an oxygen demand, using up dissolved oxygen in the water that is vital to fish and other aquatic organisms. Distillation, desalting, and other refinery processes produce small amounts of highly polluted wastewater which must be treated. Moreover, some refineries use huge volumes of cooling water which can present a serious pollution control problem if it becomes contaminated. In the past, refineries created large volumes of oil-water emulsions by cooling hot products directly with a spray of water. Modern refineries cut down on the volume of cooling water used by recycling it; and they lower the pollution potential by keeping the cooling water out of direct contact with oil and solvents. This can be done with such equipment as shell and tube heat exchangers—a small pipe containing hot product run through a larger pipe containing cooling water—and by fans which blow cooling air over many small tubes of flowing hot product.

Refineries also use steam to strip contaminants from "sour" or sulfur-containing wastewater. Steam is injected into a downward flow of wastewater, driving off hydrogen sulfide, ammonia, and light hydrocarbon gases. Stripping removes some sulfur compounds from wastewater, but the wastewater and the condensed stripping steam must be treated further. Steam is also used to strip light hydrocarbons from some fractions in the distillation tower, and to remove some gases that travel with the catalyst as it flows from the reactor to regenerator in an FCC. This too creates a small flow of highly polluted water which must be treated.

Since dirty air cannot be impounded, air pollution is usually treated at its source. Contaminated water can be impounded and treated at a central location. Water treatment systems can be primary, secondary, and tertiary. Primary systems simply separate oil and floating solids from the water. Secondary systems usually use specially grown bacteria to break down and consume oxygen-demanding, organic materials in the wastewater. Tertiary systems may include further biological treatment, physical separation, and filtration processes.

Refining produces some heavy waste solids and sludges. Some of these wastes can be incinerated, and some can be used for landfill after proper treatment. A very small volume of sludge is so toxic it must be buried.

Although the refining industry is large and essential to our economy and way of life, the public generally knows little about it. A Socal poster says, "you can't hide a refinery," but this is most true for the people who live next door. Neighbors and employees bear the brunt of pollution damage; the consumer bears the burden of control costs. The average person associates the oil business with the product she buys: the gasoline from a local dealer, the heating oil, or at furthest remove, the jet fuel that propels the plane she rides. The impact pollution control costs can have on refinery costs, profits and wholesale prices may be of financial and analytical interest, but the thing which strikes home is the impact on final customer costs and petroleum company profits. This is really the pocketbook issue: What are the final costs to the customer and the stockholder?

REFERENCES

1. Gregg Kerlin and Daniel Rabovsky, *Cracking Down: Oil Refining and Pollution Control* (New York: Council on Economic Priorities, 1975).
2. Aillen Cantrell, "Annual Refining Survey," *Oil and Gas Journal,* 1 April 1974, pp. 82-105.
3. "*Fortune's* Directory of the 500 Largest Industrial Corporations," *Fortune,* May 1974, pp. 230-255.
4. Stephen Sobotka & Company, "Economic Analysis of Proposed Effluent Guidelines, Petroleum Refining Industry, Part I," prepared for the Environmental Protection Agency, Office of Planning and Evaluation, EPA-230/1-73-020, September, 1973, p.1.
5. CEP calculation, based on: E. K. Grigsby, E. W. Mills, and D. C. Collins, "Future Capital Requirements for Refined Petroleum Products," paper presented at the National Petroleum Refiners Association Annual Meeting, 1-3 April 1973, San Antonio, Texas.
6. CEP calculation, based on data from *Survey of Current Business*, monthly issues for 1974, pp. 1, 35, 36.
7. CEP calculation, based on data in Sobotka, p. 19.
8. Federal Trade Commission and US Securities and Exchange Commission, *Quarterly Financial Report for Manufacturing Corporations,* issues for 1974.
9. CEP calculation, based on: *Statistical Abstract of the United States, 1973,* US Department of Commerce, 1973.
10. Atlantic Richfield Company, Exxon Corporation, Gulf Oil Corporation, Mobil Oil Corporation, Shell Oil Company, Standard Oil Company of California, Standard Oil Company (Indiana), and Texaco, Inc., annual reports to stockholders for 1972, 1973.
11. Sobotka; "Refinery Investment Hazards Growing," *Oil and Gas Journal,* 20 May 1974, pp. 34-35; and "Refining Capacity in US to Jump 13% by 1978," *Oil and Gas Journal,* 20 May 1974, p. 56.
12. "Trade Consequences of Environmental Cleanup," *Environmental Science and Technology,* vol. 7, no. 6, June 1973, pp. 498-499.
13. Kerlin and Rabovsky, passim, Chapters 2, 3, 4.

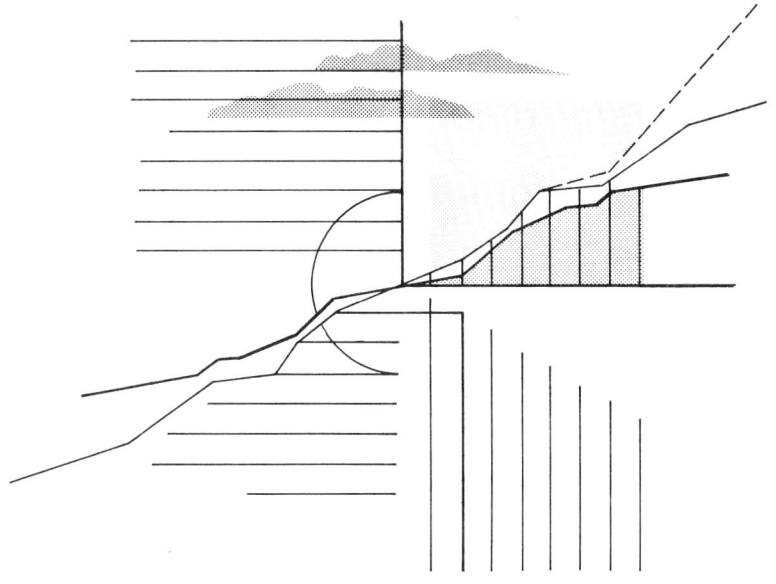

2: The Pocketbook Issue

Pollution control regulations are adding to costs in all parts of the petroleum industry. Private and government surveys indicate that only electric utilities will have to spend more for pollution control equipment over the next ten years (1,2). The added costs are huge primarily because the industry itself is huge. Considered as a per cent of total fixed assets, the pollution control capital that oil companies require is not abnormally high compared to needs in other industries.

A major portion of the integrated petroleum industry's total environmental costs goes to control refinery pollution. All by itself, the cost of refinery abatement equipment exceeds that of all US industries except electric utilities, steel and paper (see Figure 1).

Refineries have already spent a lot of money for abatement equipment. American Petroleum Institute (API) figures indicate that the industry spent at least $1.5 billion through 1972 (3). Additional large amounts were spent in 1973 and 1974 (2,4). Even so, a great deal more will be needed if the industry is to comply with the 1970 Clean Air Act and the 1977 and 1983 Water Quality Act

17

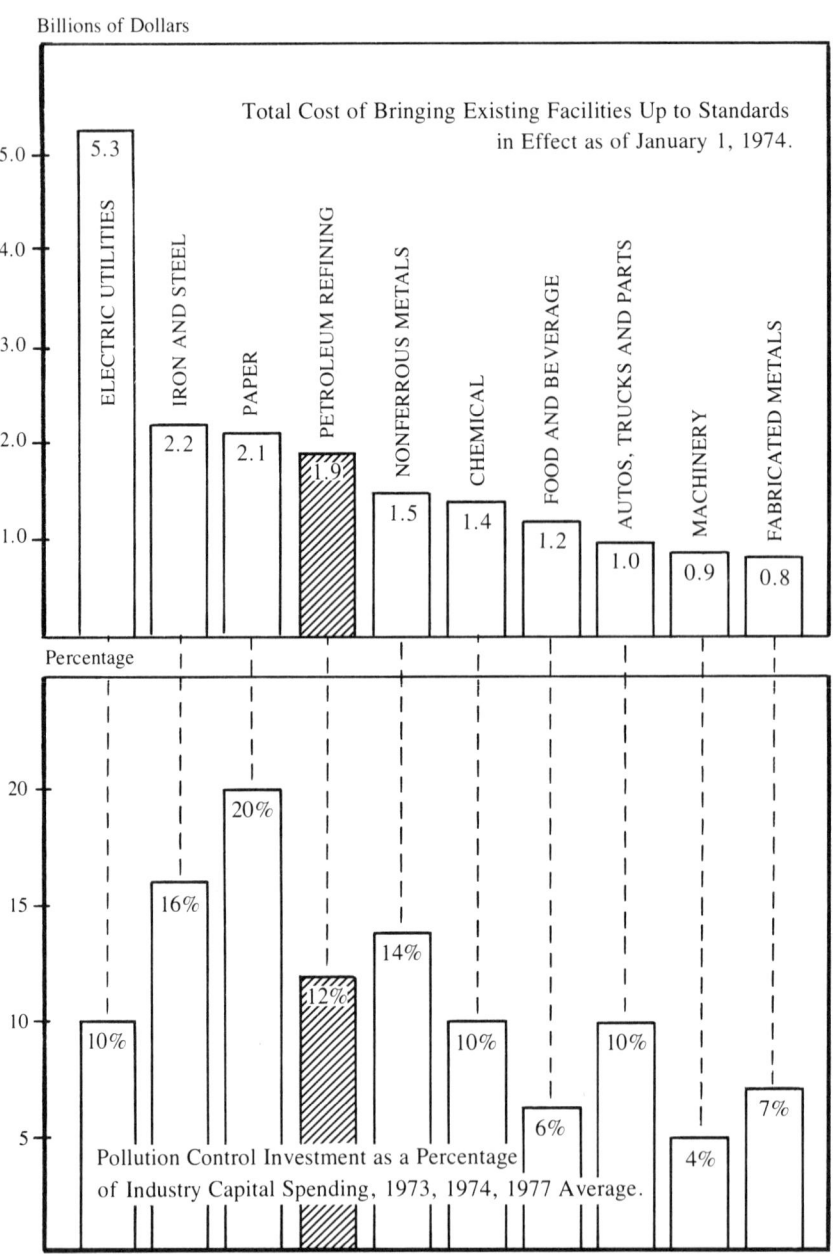

FIGURE 1. The Ten Industries that Face the Biggest Pollution Control Bills

deadlines. CEP estimates that to meet these limitations the industry will have to install $3-4 billion worth of pollution control equipment from 1974 to 1983.*

Around $2 billion of this amount is needed to bring existing refineries into compliance. The rest will be needed to control expanded capacity, assumed for purposes of estimation to grow at a 3.5% annual rate over this decade. A lower growth rate—in response to successful energy conservation, for example—would lower the total capital estimate.

Air pollution control will account for $1 billion of this total, and water abatement for the remaining $2 plus billion (5,6). Because of a combination of the past pattern of industry spending and the nature and timing of future regulations, the industry has completed much of the capital spending for air pollution control; more of the capital expense for cleaning up water pollution lies ahead (see Figure 2). In the past, refineries have concentrated on controlling air emissions. CEP findings indicate that many refiners are already close to adequate air control or have already achieved it. Two-thirds of the value of the abatement capital that was installed from 1966 to 1972 went to control air pollution (7). Severe water pollution control requirements are of much more recent vintage and the enforcement dates are a bit further off.

Capital cost estimates of this type give an indication of the size of the job to be done. It is clearly a large one, but it must be placed in context. The Chase Manhattan Bank has estimated that the petroleum industry as a whole will be spending nearly $1 trillion (1974 dollars) on capital investment over the next decade (8). The amounts needed for pollution control do not seem so huge in comparison. This industry deals in billions of dollars almost as a matter of course, and $3-4 billion does not seem to be an impossible amount of money for the industry to raise.† With respect to particular companies, CEP estimates that refinery pollution control capital is unlikely to add more than 3% to the total 1974-1983 capital spending for the worst hit firms, Shell and Arco. The amount falls below 1% for the least affected company, Exxon.

According to some estimates, 8-10% of the cost of a new, environmentally acceptable refinery would be for the pollution control equipment needed to comply with regulations (9). As we shall argue in Chapters 3 and 4, however, this does not necessarily mean that the net costs of producing petroleum products will go up by 8-10%. Much refinery pollution control equipment brings at least some return to the refiner.

---

*See Chapter 4 and the Appendix for source and details of this estimate.

†Several objections can be made to this statement. It could be pointed out that we are talking about refining pollution control alone, for US refineries at that, and other costs of control will add further huge amounts that must be financed. But environmental costs in the rest of the industry are not as large, proportionately. In addition, refining is simply too integral a part of the whole petroleum picture to separate it out for this purpose. The fundamental thing is that the parent industry has the financial base to deal with costs of this magnitude.

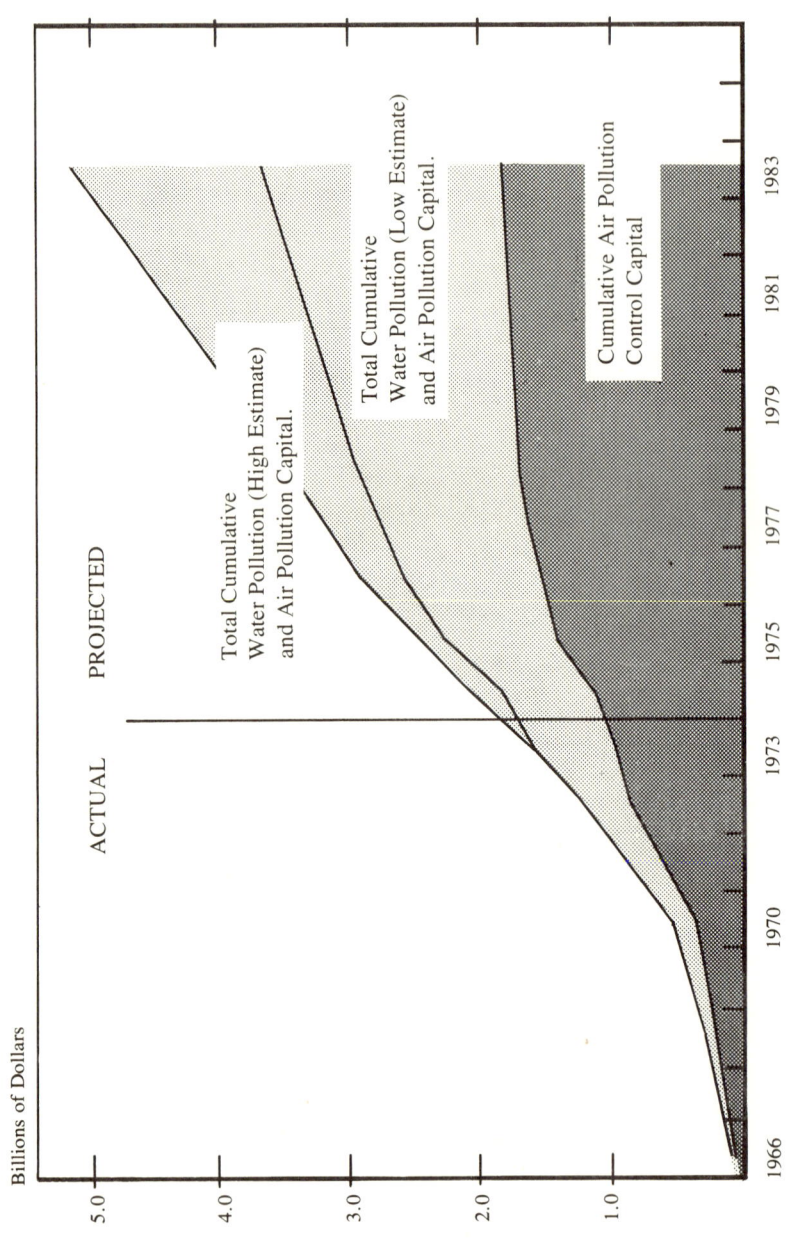

FIGURE 2. Pollution Control Capital in Place, Installed after 1965

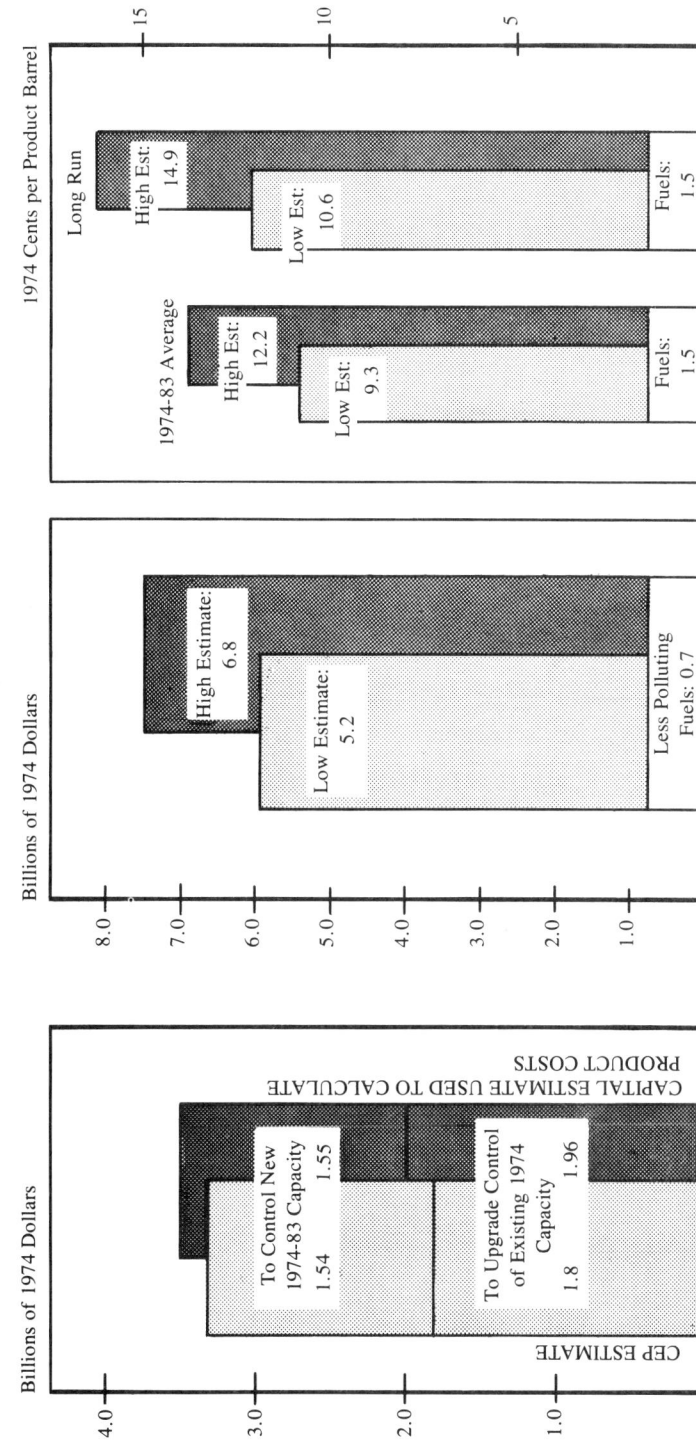

FIGURE 3. The Cost of Pollution Control, 1974-83

PRODUCT COSTS

While capital spending estimates indicate the sheer size of a job, they give us little direct information about costs and the prices of products. Increased pollution control costs will result in increased product costs because capital equipment costs money to own, operate and maintain, and because some portion of the original cost is charged against income each year (depreciation). Some additional pollution control costs are unrelated to new equipment. Less polluting fuels are a prime example.

Using the conservative model described in Chapter 4, and in detail in the Appendix, CEP estimates that the petroleum products produced from 1974 through 1983 will cost as much as $6-8 billion more to refine than they would have without improving pollution control from 1973 level. These costs come to 11-14¢ per barrel of product.*

In the long run, complying with regulations will increase product costs by 12-16¢ per barrel over 1973 levels. Figure 3 illustrates the capital and product cost estimates, both on an aggregate and a per barrel basis.

Refinery pollution control costs are a poor guide to total environmental costs in the integrated petroleum industry. Proportionally, it costs much more to control refinery pollution than it does to control pollution in other segments of the oil business. Although refineries account for only 15% of domestic petroleum assets, API figures show that refining accounted for 40-50% of total petroleum industry environmental capital spending between 1966 and 1972 (3). Accordingly, we must conclude that environmental considerations in other segments of the petroleum business as a whole will not lead to cost increases as large as those in refining, when costs are compared to business investment.

WHO PAYS THE BILL?

The added costs of cleaner refinery operations will raise product prices, reduce refinery profits, or both. Someone has to pay the bill.

This study is basically concerned with refinery costs and profits and petroleum product prices. It is inappropriate, however, to compare refinery profits directly to customer prices because so much of the customer's price is tied to the crude oil that enters the refinery—as much as 50-60% with 1974 price structures—or comes from costs that are added after the product goes out the refinery gate. These factors are reflected in total petroleum industry or company profits, and it is thus more meaningful to compare customer product (retail) prices to overall petroleum industry and company profits. Figure 4 compares

---

*The product cost figures are unaffected by the growth rate assumption. Only the total capital spending requirement is tied to this particular assumption. If new and expanded refineries are cheaper to control than already existing ones, and if the industry does grow, these figures for per barrel costs will be too high because CEP's calculation is based on the costs of completing the job at existing refineries. How much too high would depend on how much more cheaply new refineries can be controlled and how rapidly the industry expands.

those customer price increases that cover the added pollution control costs to the profit reductions the industry would experience if it absorbed the full costs of control.* Figure 5 makes a similar comparison for wholesale refined product prices and refinery profits. Each figure illustrates an extreme either/or situation.

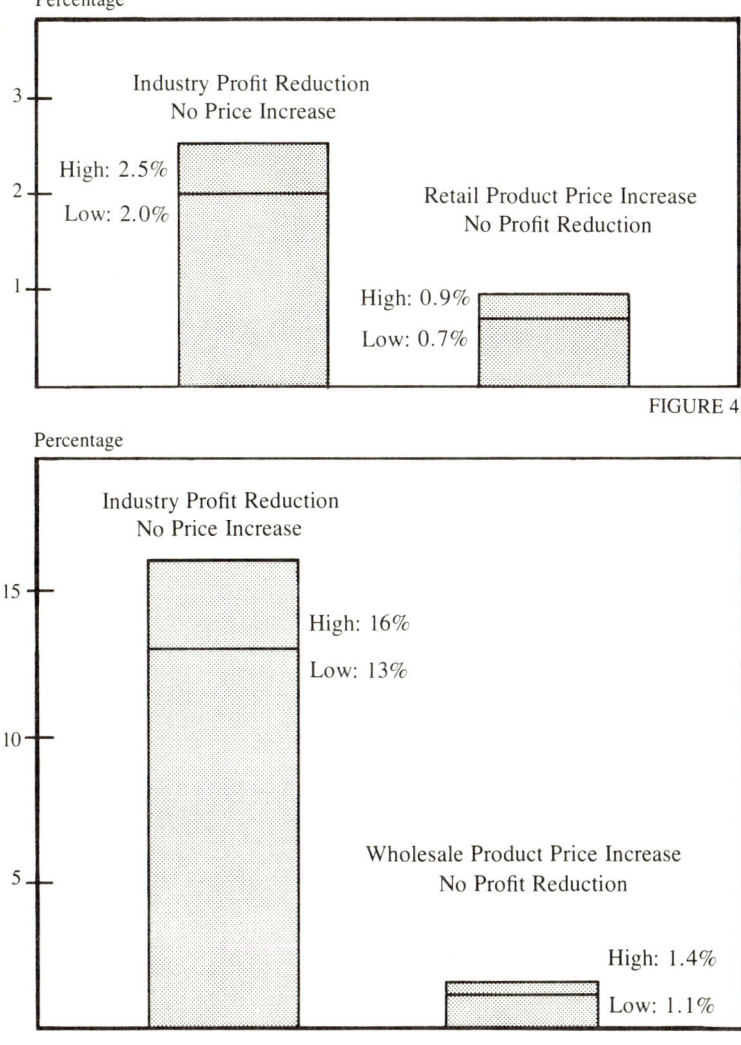

FIGURE 4.

FIGURE 5. Integrated Petroleum Industry-Profit Reductions vs. Price Increases, 1974-83

*These are industry profits, totaled over the 1974-1983 period when, we assume, profits will grow at 4% per year (1974 dollars).

WHO PAYS THE BILL? 23

In the very unlikely event that the refineries were unable to pass any of their increased costs along to their customers, profits could fall as much as 18%. For the domestic petroleum industry as a whole, this would mean a 2-3% profit reduction. Profit reductions of this magnitude would almost certainly make it difficult for the companies to obtain and justify financing for new refineries here and would lead companies to build their refineries abroad.

Many people have suggested that the petroleum industry's profits are so large it can well afford to pay for its own pollution control out of reduced profit margins. In absolute dollars, petroleum industry profits *are* huge, but so are expenses, including those for pollution control. When the profits of the past decade are compared to invested capital and to stockholder's equity, they do not appear to be outrageous compared with the returns achieved in other industries, particularly manufacturing. Over the past ten years, petroleum companies earned an average 10-11% on stockholders' investment (10). The average US manufacturing concern pulled in 11-12% in this same period. Returns on petroleum company equity certainly exceeded those of many other industries in 1973 and 1974, but it is unlikely that these profit margins will be as high in the future. A large portion of the very high profits can be traced to temporary factors such as inventory profits, and it is important to remember that the rules of the game have changed drastically in the last several years.

More changes may be coming. Oil producing countries, especially OPEC members, have largely replaced the companies as beneficiaries of high crude oil prices and profits, either through royalties, which now claim most of the sales price of a barrel of crude, or through nationalized ownership of producing wells. The nationalization of crude oil production means that petroleum companies which once owned and produced oil will now only produce. Extra profits may accrue to the producing company, but they are likely to be small compared to the windfalls of the past. Some non-OPEC countries that expect to become important oil producers in coming years, Norway and Great Britain for example, are talking about charging the oil companies very high royalties and taxes.

In the US, opportunities for large profits are also waning. US producing costs have long exceeded those of the middle East, where it currently costs less than 20¢ per barrel to bring oil from the ground. The Texas, Oklahoma and California crudes which supplied the US oil industry in the past are being replaced by offshore and Alaskan oil, which is even more costly to find and produce. Even the right to look for oil on public land costs big money. The price of federal oil leases has skyrocketed, largely to the benefit of the US Treasury and taxpayer. Put all this together with the possibility for changes in the US tax laws, in some cases aimed directly at reducing oil company profits, and it seems fairly clear that future oil profits are unlikely to be excessive.

Consequently, we conclude that refining, and its petroleum industry parent, has as legitimate a worry about the costs of pollution control as any other US business. In particular, if any industry is to compete for investment capital, it cannot stand to suffer profit margin losses which bring it returns substantially below those in other, equally risky industries.

Profit reductions equal to the full cost of pollution control are really only a theoretical possibility. Refining profits are most unlikely to be reduced by the total amount of increased cost. Instead, some of the burden will almost certainly fall on the customer. The maximum impact on prices would result if the refineries were to pass the full cost increase on to the customer. CEP estimates that the total costs of more complete refinery pollution control will amount to under 2% of 1974 wholesale product value and much less than 2% of retail value. These amounts are insignificant compared to the average 55% retail price rise for refined products during the first five months of 1974 and the earlier 18% rise from 1972 to 1973.

A price rise equal to the cost increase would maintain refinery return on investment and total profits only if product sales did not fall at all as prices rose. In that case an average 1% increase in retail prices would recover all pollution control costs for the refiner and maintain return on refinery investment at its current level. If product demand fell when refiners raised prices, a price rise greater than the unit product cost increase would be needed to generate enough added revenue to equal the full cost increase.

Assume, to take an extreme example, that full cost recovery is possible and that it comes solely through price rises for gasoline, which comprises 45-50% of refinery production. Prices for all other refinery products remain unchanged. Assume further that a 10% rise in retail prices causes a 1% decline in the amount of gasoline sold. To recover costs completely, then, prices would have to rise by 112% of the estimated cost per barrel. This comes to 1¢ per gallon of gas, a 2% increase on a 50¢ retail gallon. Spread over ten years, this price increase amounts to an insignificant 0.2% per year. A 2% demand decline stemming from the 10% price rise could be recovered by a 2.5% price rise. These are outer limit price increases because prices of the other 50% of refinery products would also rise, and the full costs could be recovered with significantly smaller gasoline price rises.*

The strength of demand will largely determine the way the added costs of pollution control will be shared between the general customer and the refinery. Historically, the demand for refinery products has fallen very little in response to petroleum product price increases (11). Consumers may become more resistant to higher prices, however, as they reach levels unprecedented in the US. Alternative modes of transportation, for example, may become increasingly attractive, and public demand for mass transit will probably rise. Thus, while the marketplace absorbed very large petroleum price increases in 1973-74 with little immediate reduction in demand, it has become clear that at some level, product purchases are affected by price increases. The Ford administration's economic

*The illustrative figures used to indicate the impact of a price rise on product demand are probably of the right order of magnitude even though the actual degree of responsiveness of demand for petroleum products to price changes is not known with any accuracy. Prices in this industry have fluctuated so little in the past, and they have been at present levels for such a short time, that very little research has been done on this matter.

advisors clearly expect demand to fall as prices rise and have therefore recommended taxes on petroleum products which will significantly increase their final prices.

Almost certainly the customer will pay the lion's share of the increased costs that result from pollution control. The demand for refined products is still strong and there are few, if any, significant substitutes for them. So long as this is the case, refiners will be able to pass much of the pollution control burden along to their customers.

Figure 6 very roughly indicates the possible combination of price increases and petroleum industry and refining industry profit reductions. For purposes of estimating profit impacts, CEP assumes that product price rises will enable refineries to recover 75% of the cost increases. A 75% recovery means that retail product prices would rise by about 1%, assuming little reduction in demand, and the average petroleum company would suffer approximately 1% profit reductions. Seventy-five per cent is a somewhat arbitrary figure, but its order of magnitude is not unrealistically optimistic for the refiners.

At a 75% cost recovery level, neither the public nor the refining and petroleum industries appear to be paying an impossible price. For the industry in particular, pollution control costs should not render rates of return significantly lower than returns in other businesses. After all, other industries face the same problem, and many of them are by no means as able to pass along cost increases.

A 100% cost pass-through may be possible, but it is far from certain. This industry is highly capital intensive and has very large fixed costs. Price resistance may reduce sales, causing refiners to cut production. Other costs per unit of product would then rise, and prices would have to rise even further. Demand would fall more and so on. At some point, the industry realizes the highest return it can get on its investment. That is not necessarily a return which achieves the level attained before the added costs came along. Only a careful and lengthy analysis of the industry cost structure could determine the significance of this factor.

It must be emphasized that all of this is in terms of petroleum industry averages. Petroleum products are largely undifferentiated within class and type except by price. Gasoline is a case in point. Oil companies frequently trade gasoline to supply markets more easily, simply adding their own special compounds to the final mix. One company is not able to sell its products for consistently higher prices than its direct competitors. Added 1974-1983 pollution control costs per barrel of product, however, will differ from firm to firm, depending on how much pollution control has already been achieved and perhaps, on proportional rates of expansion if it is cheaper to control new refineries. If price increases are about the same throughout the industry, different firms will recover different percentages of costs. Conceivably, one firm could actually recover more than its cost increase and come out ahead on the deal. CEP's cost analysis indicates this *may* be the case for Arco and Shell if the industry recovers 75% of average cost (see Chapter 5 for a further discussion of this matter).

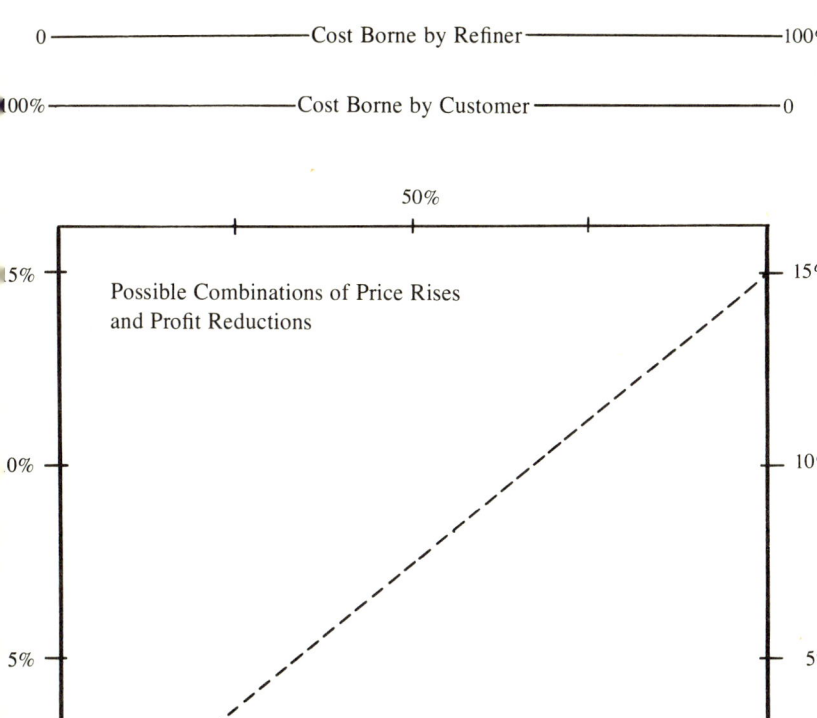

FIGURE 6. Who Pays the Bill?

Key

Reduction in Refinery Profits

Increase in Wholesale Product Prices
—·——·——·——·——·——·——·——·—··—

Reductions in Petroleum Industry Profits
————————————————

Increase in Retail Product Price
———·———·———·———

There is one more bearer of the pollution control cost burden—the US taxpayer. Even if prices do not increase at all, under US tax law, the general taxpayer, including corporations, foots approximately half the pollution control bill. This is because businesses deduct costs of operation from their revenue to calculate net income. If the 48% marginal income tax rate (the tax rate on the final dollar of income) applies, refiners will bear only about half the $6-8 billion cost in terms of reduced profits. The taxpayers' burden will fall if prices rise. Some portion of increased revenues go to the government in the form of income taxes. If cost recovery is total, the taxpayer's bill will be zero since higher taxes from increased revenues just offset lower taxes from increased costs.

The tax consideration thus means the public (here meaning everyone including the petroleum companies themselves) will pay from half to all the control bill. Even if the whole cost is passed along to the customer, however, the public is not necessarily out of pocket on a net basis. Refinery pollution imposes

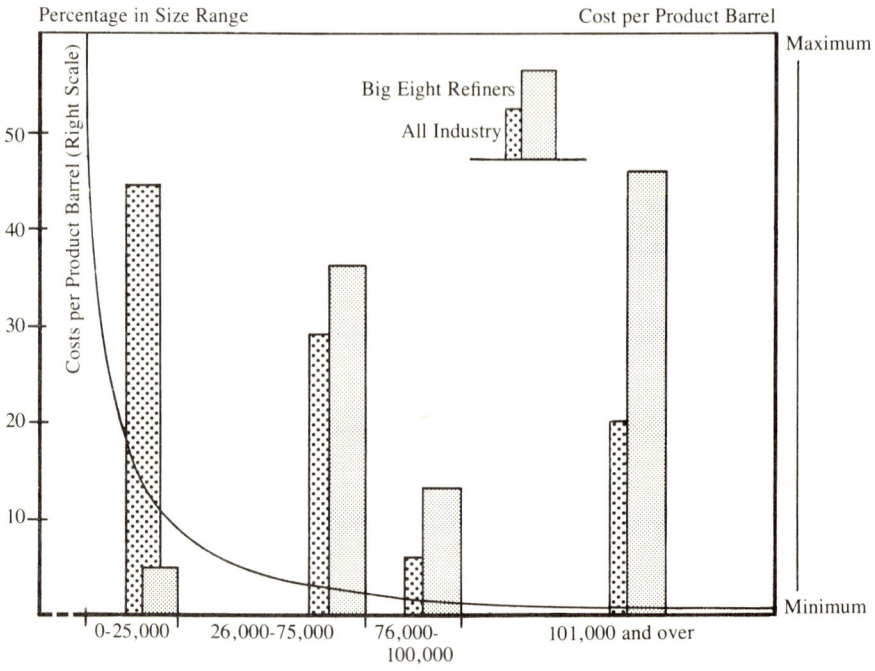

FIGURE 7. Refiners Distributed According to Capacity and Control Costs

28 THE POCKETBOOK ISSUE

substantial costs on the population in the form of impaired health, damaged agriculture and diminished recreation facilities, to cite just a few examples. We pay for pollution one way or another. So far, none of the attempts to quantify the damage pollution does has been entirely satisfactory, and none is specific to the pollution produced by petroleum refining. One tentative attempt to assess air pollution damage from refining suggests that such damage probably exceeds the costs of control by a wide margin (12).* If this is so in general, the public as a whole gains even if price rises cover the full amount of the clean-up cost.

NOBODY'S AVERAGE

No two refineries are exactly alike in their engineering design, product composition, or capacity. The figures CEP presents in this report are averages which do not fit any refinery exactly. Pollution control costs per barrel of product are not the same for all sizes, all complexities, or all product configurations. They tend to rise with greater complexity and fall with larger size. When attempting to apply the results of this study to individual situations, remember, the suit may not fit without extensive alterations.

The differences in refinery costs that arise from varying capacities illustrate the distortion potential of average figures. Refinery capacities range from the very small, under 2,000 bbl/d, to the giants that process over 400,000 bbl/d (11). The industry average is 58,000 bbl/d. (The study average is 130,000 bbl/d.) Engineering studies indicate that unit product control costs are very high for small refineries and fall rapidly as size increases to 25,000 bbl/d (13). Cost continues to fall somewhat more slowly up to 75,000-100,000 bbl/d and then levels off. Average costs applied to units with capacities above 25,000 bbl/d probably do not result in any important distortion because of size considerations. A large company's refineries tend to be larger than those of the industry in general, suggesting that unit control costs for the big eight will tend to be lower than the industry average. Figure 7 shows the size distribution for both the industry and the big eight refiners.

The costs of controlling refinery pollution should have an uneven impact on the profits of the eight firms covered in this study. Current levels of pollution control at existing capacity differ from firm to firm. Those companies which have completed more of the job will bear lower added costs in the future than the companies that have much left to do. Companies with higher than average expansion rates are in an advantageous position since it may be cheaper to build

*Lyndon Babcock, Jr. and Niren Nagda published a study, "Cost Effectiveness of Emission Control," in March, 1973 (12). The study allocated pollution damage costs to industrial and private sources of air pollution and estimated how cost effective pollution control would be for each of the sources. They found that air pollution control at refineries would reduce air pollution damage by $260 million per year. This amount was clearly in excess of estimated annual costs for air pollution control. Indeed, they estimate that air pollution cost effectiveness is as high, or higher, for refineries than for any other major source of air pollution. The study, however, was only a tentative attempt to quantify pollution damage costs and did not represent a careful investigation. Hence we regard the results as only suggestive.

control into new capacity than it is to retrofit. Finally, in a total company context, the firms have invested varying proportions of their total capital in domestic refining.

CEP estimates that the 1974-83 costs of complying with pollution control regulations will reduce profits most for Gulf and Amoco and least for Exxon and Arco. Table 4 supplies more specific estimates for all eight companies.

Pollution control costs money, for the public which buys products, for the taxpayer, and for the petroleum company. These costs, however, are not fixed, unchanging or clear cut. They are frequently misunderstood and often overstated. The real, and changing, costs of pollution control are the subject of the next chapter.

TABLE 4

Projected Reduction in 1974-83 Profits due to Pollution Control Regulations

Company	Zero Cost Recovery	75% Industry Average Cost Recovery
Atlantic Richfield	3.5–4.5%	.2– .3%
Exxon	.7– .9%	.1%
Gulf	2.1–2.7%	.8–1.1%
Mobil	2.0–2.4%	.7– .9%
Shell	4.6–5.7%	.5– .6%
Standard Oil of California	2.5–3.2%	.7– .9%
Standard Oil (Indiana)	3.3–4.1%	.8–1.0%
Texaco	1.6–2.0%	.5– .7%

REFERENCES

1. John E. Cremens, "Capital Expenditures by Business for Air and Water Pollution Abatement, 1973 and Planned 1974," *Survey of Current Business,* July 1974, pp. 58-64.
2. McGraw-Hill, Seventh Annual Survey of Pollution Control Expenditures, reported in *Air/Water Pollution Report,* 27 May 1974, pp. 202-203.
3. CEP calculation, based on: American Petroleum Institute, "Environmental Expenditures of the US Petroleum Industry, 1966-1972," API Publication #4176, 1973. See Appendix for method of calculation.
4. Gregg Kerlin and Daniel Rabovsky, *Cracking Down: Oil Refining and Pollution Control* (New York: Council on Economic Priorities, 1975), passim.
5. Stephen Sobotka & Company, "The Impact of Costs Associated with New Environmental Standards upon the Petroleum Refining Industry," prepared for the Council on Environmental Quality, November, 1971; and Sobotka, "Economic Analysis of Proposed Effluent Guidelines, Petroleum Refining Industry, Part I," prepared for the US Environmental Protection Agency, Office of Planning and Evaluation, September, 1973.
6. Brown & Root, Inc., "Economics of Refinery Wastewater Treatment," prepared for the American Petroleum Institute, Committee on Economic Affairs, API Publication #4199, August 1973.
7. American Petroleum Institute, "Environmental Expenditures of the US Petroleum Industry, 1966-1972," API Publication #4176, 1973, Appendix I, Tabulation of Survey Data.
8. Richard S. Dobias, Norma J. Anderson, Richard C. Sparling, with John C. Winger, "1973 Annual Financial Analysis of a Group of Petroleum Companies," The Chase Manhattan Bank, Energy Economics Division, October 1974.
9. W. L. Nelson, "Questions on the Technology, Refinery Pollution Control," *Oil and Gas Journal,* 4 September 1972, p. 86.
10. Sobotka, 1973, exhibit 15.
11. Sobotka, 1971, p. 7.
12. Lyndon R. Babcock, Jr., and Niren L. Nagda, "Cost Effectiveness of Emission Control," *Journal of the Air Pollution Control Association,* vol. 23, no. 3, March 1973, pp. 173-179.
13. Environmental Protection Agency, "Economic Analysis of Proposed Effluent Guidelines, Petroleum Refining Industry, Part II," prepared by the EPA Office of Planning and Evaluation, September, 1973, pp. 23-28.

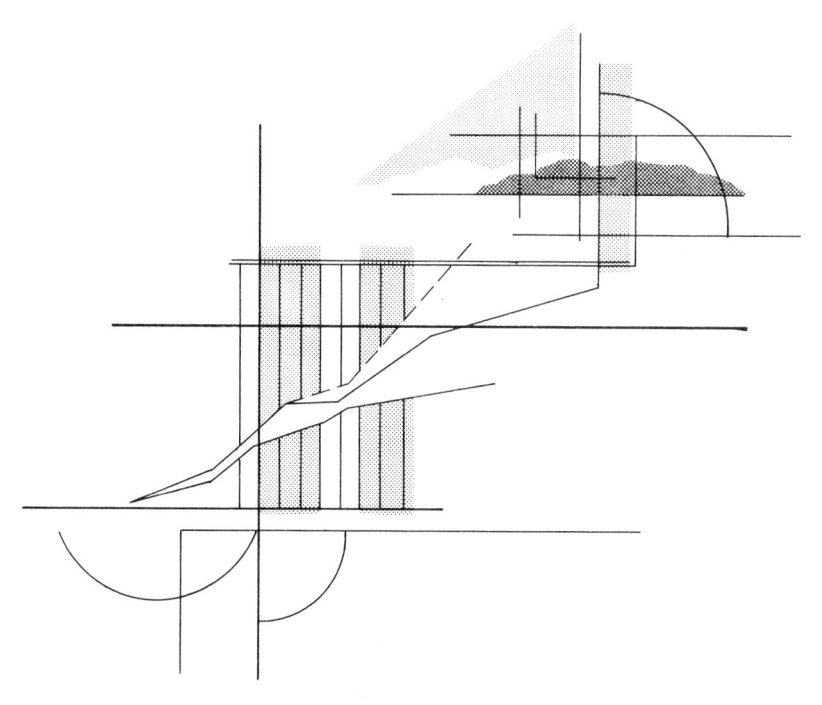

3: What *Are* Pollution Control Costs?

Any analysis of the costs of pollution control suggests that it is legitimate to consider pollution control in the abstract, as a cost of doing business separate from ordinary operations. This is appropriate only from certain points of view. A company might state pollution control costs separately in the hope of convincing the government and the public that the costs are high enough to justify looser regulations. The public should be interested in separately stated costs to determine whether the added costs are greater or less than the known or assumed costs of pollution damage. Stock and bond investors want to know the amount of costs added during the period of transition from no or limited control to full compliance since profits may be affected by previously unrequired costs.

Once environmental regulations are in force, however, it is no longer appropriate for the polluter to consider pollution control costs separately from other costs of operation. The costs of employee washrooms, for example, are never stated separately from the costs of refinery construction because they are considered a normal and necessary part of the construction. It would be impossible to get workers without them. Pollution control should be no different. It too is a necessary cost of doing business—the refinery will be fined or even shut down unless the controls are installed. (In this sense, pollution control has a potentially tremendous rate of return!) Rather than calculating rates of return on refinery investment and then moaning that estimated additional pollution control costs will reduce these rates, good investment analysis includes abatement capital in the original investment package.

None of this denies that pollution control costs money. Control frequently adds to the costs of production (although not necessarily to the total costs of the economy since monies are saved as pollution damage is reduced). Since regulations have not been fully implemented or accepted yet, it still seems appropriate to consider these costs.

CHEAPER THAN IT LOOKS

Pollution control expenditures increase production costs only to the extent that control brings no offsetting returns to the polluter. The polluter can realize returns in several ways:
1. Use or reuse recovered pollutants
2. Sell recovered pollutants
3. Reduce the damage the pollution does to the polluter's own production process, including harm to employee health and morale.

In a pure case, the gross costs of abatement would equal net costs, the recovered pollutant would have no further value to anyone, the polluter would suffer no ill effects from its own pollution, and so gain nothing from abatement. Such a pure case is probably nonexistent. Pollution control is almost always cheaper than it looks.

The relation between the cost of abating pollution and the polluter's returns from abatement are crucial factors in calculating net pollution control costs. Polluters pollute because the costs of abatement exceed the value of the returns. It is important to remember, however, that "value" is really an economic and technological term, a concept with no absolute meaning. Everything has a use at some set of relative prices and with some, perhaps undiscovered, technology. Change prices and technology, and the whole calculus of pollution control costs will be upset. This creates a problem for cost analysis because relative prices, technology, and therefore the potential returns to the polluter from pollution control change over time.

The refining industry is an outstanding example of a business which can gain from pollution control. At present, efforts to clean air offer much greater returns to refineries than water pollution control does. The most valuable refinery pollution control by-products are the sulfur recovered from hydrogen sulfide emissions and the hydrocarbons, both crude oil and products, recovered by floating roof tanks. API separators, which remove crude oil from wastewater, are the only major piece of refinery water control capital with an important return.

Returns from these separators are so good that many authorities do not even list them as pollution control equipment.* Research into water pollution control and its by-products may soon add to the returns from cleaning water, turning last year's water pollution problem into next year's moneymaker.

Refining is replete with examples of pollutants that have become products and processes that have become profitable. The whole petrochemical industry is a case in point. Similarly, kerosene was the prime refinery product in the late 19th century. There was little demand for kerosene's natural companion, gasoline, until the internal combustion engine became common. At times, the excess supply of unsalable gasoline was so great that refineries simply had to dump it to dispose of it. Alternative disposal methods would have been titled "pollution control." Today, gasoline constitutes 40-50% of refinery product output by volume and more in terms of value. Pure kerosene has become a minor product, except as a component of jet fuel.

Although they are not regarded as primary products, some refinery pollutants can be recovered and sold to outside customers. The elemental sulfur and sulfuric acid that sulfur recovery plants produce from refinery emissions are in this class. EPA figures indicate that the cost of sulfur recovery plants comprise approximately 20% of total air pollution capital for a large refinery (2). At the present time, operations for a single, three-stage Claus plant have an acceptable investment rate of return at any sulfur price above $40 per ton (3). Sulfur prices have been at this level and higher in the past year, but historically, the price of sulfur has been volatile. Some refineries were installing sulfur recovery plants for investment purposes as long ago as the 1950's, but the variability of sulfur prices makes such an investment risky. Back-up sulfur plants may be assumed to be strictly for control purposes, with little or no return (4).

Installing floating roofs on crude and product storage tanks can prevent evaporation, almost eliminating hydrocarbon emissions from that source. Recent dramatic increases in crude oil and product prices have so increased the return from recovering these emissions that many refiners find this form of pollution control an attractive investment. Shell told CEP, "floating roof tanks . . . have become economically justifiable in most cases due to recent increases in the value of petroleum products (5)." The API itself assigns only 20% of the cost of new floating roof tank construction to pollution control (1). As of 1973, however, the API still considered 100% of the cost of converting fixed roof tanks to be pollution abatement costs. At 1974 crude and product prices, even the 20% assignment seems excessive. When the full cost of floating roof tanks is included in a company's pollution control bill, the potential for overstating net pollution control costs is sizable. According to EPA figures, this type of equipment could amount to nearly 20% of capital investment for a large refiner's air pollution

---

*Even the American Petroleum Institute (API), the industry trade organization which originally developed these separators, conspicuously omits them from its list of water pollution control equipment (1).

The use of carbon monoxide (CO) boilers provides a good example of recycling pollutants within a refinery to make partners of energy conservation and pollution control. These boilers burn carbon monoxide wastes from the fluid catalytic cracker, producing heat which can make steam to power other refinery processes. The boilers are usually listed as air pollution control equipment, although they were originally installed as a normal part of the fluid catalytic cracking process, not specifically to control pollution. At present fuel costs, this equipment generally has a positive rate of return, sometimes an attractive one. Charged at full cost, these boilers would account for more than 30% of air pollution control capital cost (2). The API currently assigns only 50% of CO boiler capital to pollution control (1), but even this is an overstatement as long as energy and fuel prices remain high.

One final example should be mentioned since it is intimately related to energy questions. Refinery processing gives off significant amounts of by-product gases. Refiners have found that these gases are one of the cheapest sources of fuel they can use despite the special refinery design that is required to capture and reuse them. When the gases are not used for fuel, they are usually flared off, creating pollution and increasing the amount of outside energy the refinery must use.

This by no means exhausts the list of ways in which refineries can recycle and reuse potential pollutants. Socal sells recovered ammonia to agricultural fertilizer manufacturers (6); Exxon is investigating the use of treatment plant sludges as sanitary landfill (7); phenols and other pollutants may also have uses. Most refinery catalysts are recovered and reused because they are so expensive. In fact, this was the original purpose of cyclones. The catalyst particles that escape to become pollutants are those which are too small to be recovered economically. Other examples of reuse include spent caustic, which is regenerated rather than released to pollute the environment.

OVERSTATED COSTS

Many refinery treatment and separation operations flirt with the fine line between pollution control with a maintenance by-product and maintenance that results in pollution control. Cutting hydrogen sulfide emissions can reduce the maintenance expenditures hydrogen sulfide corrosion causes. In addition, it can result in significant revenues from the sale of recovered sulfur. Careful maintenance and process control often significantly reduce water pollution flows and cut down the amount of crude and product lost through spillage. Pollution control in situations such as this saves on the ultimate costs of production.

Labor is a critical input to the production process, and people provide that labor. If pollution control at the plant results in better employee health and morale, the refinery's productivity is likely to increase. This may be a cold-blooded point of view—employee health has value independent of productive factors, but it points up the fact that the employer has an important economic interest in the well-being of his employees. This return from control is subtle, but it is not necessarily trivial.

A study of pollution control in the pulp and paper industry found that, contrary to conventional wisdom, above-average profits often went along with good pollution control practices (8). This did not appear to be because the more affluent firms could better afford to be clean. Good management was cited as a key factor in a company's decision to institute good pollution control, but once that decision had been made, several subtle cost offsets were suggested as reasons for the higher profits of better controlled companies. These included improved employee health and morale, resulting in increased productivity, reduced labor costs, and lower labor turnover; lower financing costs for the cleaner company if the investment community thinks the firm has a better image or takes pollution control performance as a measure of the ability and foresightedness of management; lower pollution control costs when abatement is part of original plant design rather than retrofitted; and higher company revenues if customers prefer the cleaner company.

Most, if not all, of the types of returns cited above are difficult to quantify. Many analysts do not even try, simply ignoring all but the most obvious. The result is an overstatement of the true net costs of control. For example, most analysts would say that the cost of equipment a refiner would install in the absence of pollution control considerations, or at least in the absence of government regulations, includes no pollution control component no matter how good a control job the equipment does. Alternatively, the full cost of equipment that is installed in response to regulation is commonly assigned to pollution control.

As we have seen, ignoring the returns that control brings to the polluter can, in some cases, overstate those costs which could reduce profits by as much as 30-40%. Businesses generally determine a minimum acceptable rate of return for investment, say, a projected 15% after tax. A pollution control project, such as a sulfur recovery plant, could have a positive rate of return which is under that minimum, and so it would not be undertaken without the push of regulation. The company would be likely to cite the full cost of the project as pollution control although revenues are generated and at least some portion of the invested capital has an acceptable return.

Inadequate accounting for returns is only one reason that pollution control costs are frequently overstated. Another has to do with the political environment in which business operates. Pollution control is a "good" in the public mind, and rational businessmen want to get as much public relations mileage as possible from pollution control expenditures. Emphasis and overstatement is a natural result. In an earlier report, "Corporate Advertising and the Environment," CEP found that it is not uncommon for businesses to exaggerate environmental expenditures in public relations material (9). Even when a company attempts to allow for returns and carefully separates cost elements into pollution control and productive categories, the allowance can often be no more than a seat of the pants affair. Pollution control is unlikely to be the understated cost element (10).

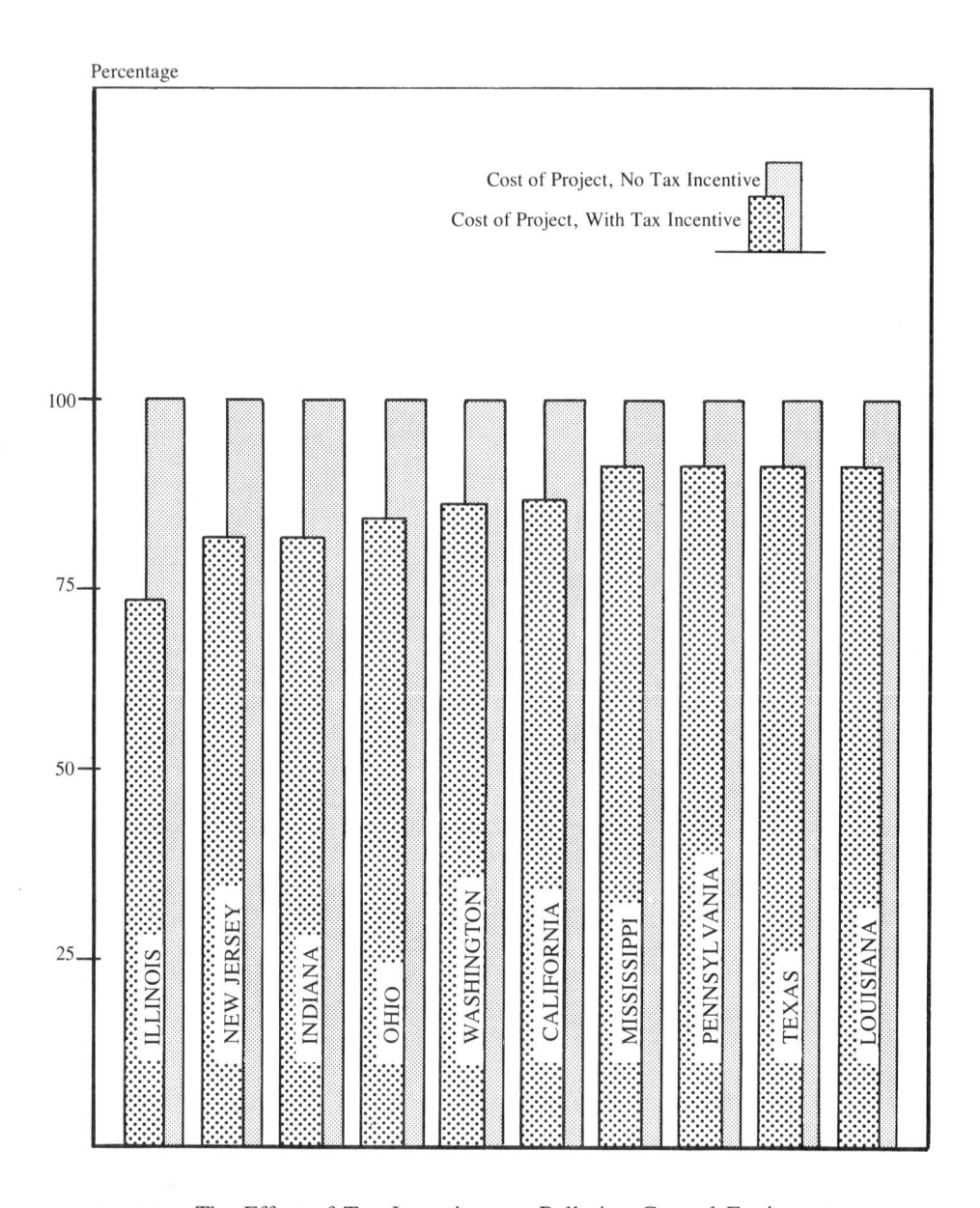

FIGURE 8. The Effect of Tax Incentives on Pollution Control Equipment Costs in 10 Important Petroleum Refining States

Federal and state tax codes also encourage a stress on pollution control costs. Both federal and state governments offer incentives and tax subsidies to encourage the installation of pollution control equipment. All of these reduce the investment cost of pollution control equipment compared to the costs of production equipment. In effect, the taxpaying public pays part of the pollution control bill. Revenues that are lost to government when corporations use pollution control tax credits, exemptions and subsidies must be made up from the general tax fund. The cheaper pollution control financing is, compared to other investment, the greater the incentive to classify investments as pollution control.

At the present time, the cheaper interest cost, tax-exempt revenue bonds that may be used to finance pollution control capital are the most generally important special pollution control aid (11). In addition, the federal government offers an optional accelerated depreciation, some states offer property tax exemptions, and a few jurisdictions go so far as to provide direct subsidies.

These subsidies have a large potential value. A study of the tax incentives available for pollution control (11) found that government tax deals could significantly reduce the capital cost of pollution control facilities. The most important single vehicle was tax exempt bond finance. Present value cost reductions for significant petroleum refining stages ranged from a 27% reduction in Illinois to 9% in Louisiana. The Louisiana reduction was due almost entirely to the use of tax exempt finance. Figure 8 indicates the present value cost of pollution control capital compared to the present value cost of ordinary business investment for ten important petroleum refining states. The eight companies in this study, for example, have nearly 90% of their refining capacity in these ten states.

It is extremely unlikely that every polluting company could achieve such cost reductions for their pollution control capital. In particular, if every nominally eligible corporation were to use tax exempt bond finance, the tax exempt market would be overwhelmed. In this case, Congress might well repeal the enabling legislation to preserve the market for state and local government bond finance. At the moment, however, pollution bonds are being floated and petroleum companies are among those issuing them. A prime example is the $110 million pollution abatement bond Exxon issued in April 1974 and earmarked for pollution control at refineries and chemical plants at Baton Rouge, Louisiana (12,13). The company's interest cost may have been reduced by as much as one-third by using this vehicle.

Most of the other important tax subsidies and incentives are provided by state and local governments. The federal government's accelerated depreciation is of questionable value. Very few pollution projects are actually eligible for it, and those that are lose the investment tax credit for the period in which they elect the accelerated depreciation (14).

Companies that need pollution control capital can also benefit from the tax breaks that are available to capital equipment in general. The investment tax credit, worth a tax reduction equal to 7% of the cost of the equipment, is a major example. CEP estimates for product costs assume that the refining industry and the companies take full advantage of this tax credit. This is the only special tax provision incorporated into the estimates.*

CEP'S COST ESTIMATES

The cost results reported in Chapter 2 are almost certainly too high in net terms. They should be regarded as an extreme upper limit only. CEP's cost estimates are plagued with the problems of overstatement that are endemic to this industry. Without our own engineering and capital budgeting study of refineries, we have had to rely to a great extent on other people's figures for capital investment needs. Returns from control are almost certainly inadequately accounted for, although the air control capital estimates do make some allowances. The next chapter describes how we arrived at our cost estimates.

*Proposals to change the nature and value of the credit have been submitted to Congress. Whether the new version would raise or lower the actual value seems to be unclear, and until the final form is determined, it seems best to use the old form for analysis.

REFERENCES

1. American Petroleum Institute, "Environmental Expenditures of the US Petroleum Industry, 1966-1972," API Publication #4176, 1973, Appendix II, the Survey Questionnaire, list of air and water pollution equipment.
2. CEP calculation, based on: Environmental Protection Agency, "The Economics of Clean Air," Annual Report of the Administrator of the Environmental Protection Agency to the Congress of the United States, March 1972, p. 4-111.
3. Matt Noble, Exxon USA, personal communication, 19 July 1974.
4. Gregg Kerlin and Daniel Rabovsky, *Cracking Down: Oil Refining and Pollution Control* (New York: Council on Economic Priorities, 1975), Chapter 5.
5. Shell Oil Company, response to CEP economic and financial questionnaire.
6. Standard Oil Company of California, response to CEP economic and financial questionnaire.
7. Exxon Corporation, response to CEP economic and financial questionnaire.
8. Joseph H. Bragdon, Jr., and John Marlin, "Is Pollution Profitable?," *Risk Management,* April 1972, pp. 9-18.
9. Leslie Allan, David Johnston, Joanna Underwood, and Ranne Warner, "Corporate Advertising and the Environment," *The CEP Economic Priorities Report,* vol. 2, No. 3, September-October, 1971.
10. W. L. Nelson, "Costs of Refineries, Part 4," *The Oil and Gas Journal,* 29 July 1974, p. 162.
11. Thomas P. Broderick, "An Analysis of Tax Incentives for Certified Pollution Control Facilities," unpublished Masters essay, University of California, Berkeley, 1974, p. 59.
12. Richard R. Leger, "More Companies See Tax-Exempt Bonds for Pollution Control, Saving Millions," *Wall Street Journal,* 8 July 1974.
13. Bond Prospectus, dated April 23, 1974. Parish of East Baton Rouge, Louisiana: $110,000,000 Pollution Control Industrial Revenue Bonds, 5.90%, Series A, Due May 1, 1999. Payment of the principal and premium, if any, and interest on the bonds is guaranteed by Exxon Corporation.
14. Broderick, p. 51-53.

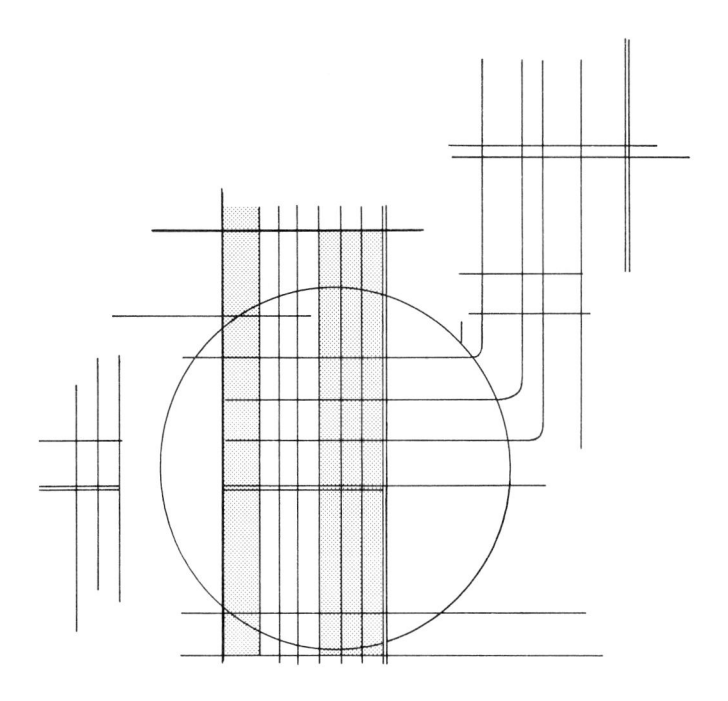

4: Calculating Costs

Controlling refinery pollution is a complicated and potentially expensive business. Control equipment must be installed, then operated and maintained. The refiner may have to alter refining processes, use cleaner, more expensive fuels, and tighten up general maintenance and cleaning practices. All this can add up to a very large bill, even after account is taken of the returns. Before we go on to consider the size and types of these costs, however, it is necessary to dispose of the issue of the polluting product, which is a red herring in the context of this discussion.

The nation's refineries face two distinct types of added environmental protection costs. They must clean up and control the pollution their own production processes cause, and they must make less polluting products, such as unleaded gasoline, which cut down the pollution produced by customer product use. The industry often includes both these costs in figures for environmental protection. The combined dollar figure is high, and it vastly overstates the amount refiners must spend to control their own pollution. The equipment that produces clean products is very costly. Industry estimates suggest that as much as $5-6 billion worth of extra equipment will be needed to meet the anticipated demand for low sulfur fuels and unleaded gasoline (1). While the monies needed to install and control this new equipment are certainly legitimate pollution control expenditures in the context of the whole economy, they are not *refinery* pollution control costs.

The industry should be able to recover most of the added costs of making environmentally acceptable products by raising prices. If the additional costs cannot be recovered fully, refiners will be stuck with reduced profit margins on products. In that event, the government will have forced them to produce a less profitable line of products and to absorb some of the customer's pollution control costs. Again, this legitimate pollution control cost would not be a *refinery* cost.

Producing clean products can affect refinery pollution control costs, not because they cost more to make, but because the amount of pollution a refinery produces and the effort it takes to control it is affected by the products it produces. More thorough refining is needed to make cleaner products, and the new processes sometimes increase a refinery's air pollution potential. The real pollution costs at the refinery need not rise, however, because it is often easier or cheaper to control pollution from these new sources than it is to control older processes. Some of the more thorough processes may actually reduce water pollution potential.

EQUIPMENT COSTS

Clean refining costs plenty all by itself. The costs fall into two main categories: costs related to capital installation and costs connected with changes in operating procedures and more meticulous housekeeping. Capital linked costs are by far the most important.

Pollution control investment can be for equipment which is specifically intended to control pollution and has little relationship to the production process, or it can be for environmentally motivated changes in the production process. If a refinery engineer chooses a cleaner, more expensive production process over a cheaper, dirtier alternative, a genuine pollution control capital element exists, even though there is no pollution control investment as such. Choosing air cooling towers, which greatly reduce refinery water use and therefore water pollution, in preference to water cooling methods is an example of this type of capital cost.

Both public and private authorities have estimated the capital amounts refiners will need to comply with government pollution control regulations. The estimates, made by projection, imputation, engineering studies, and just plain wishful thinking, generally attempt to cover both specific control equipment and the added costs of more expensive production methods.

The EPA and its contractors, Roy F. Weston and Stephen Sobotka & Co., have estimated the costs of both air and water control (2,3,4). Brown & Root,

Inc. (5), working under contract to the API, has given figures for the capital costs of water pollution control. API surveys have reported actual capital expenditures for the years 1966 through 1972 (6), and the United States Commerce Department has provided figures for 1973 (7) as well as limited projections for the future. CEP has made an additional rough estimate of its own to serve as a check on the previously published figures. Table 5 summarizes all of these capital investment estimates.

TABLE 5

Estimated Pollution Control Capital Investment
Needs for the Refining Industry, 1974-1983.

(All figures are in millions of 1974 dollars.)

	CEP	EPA	API
1974-1977			
Air		640	640*
Water		1,000	1,150
TOTAL	1,440	1,640	1,790
1978-1983			
Air		300	300*
Water		1,000	1,925
TOTAL	1,900	1,300	2,225
1974-1983 TOTAL	3,350	2,940	4,015

*No estimate other than that of EPA.

The figures listed in Table 5 for the EPA and the API differ from numbers found in the original sources because CEP has converted all these numbers to a common basis of 1974 dollars, an assumed 3.5% annual industry growth rate, and a 1973 base for pollution control capital stock.

The industry has grown at an average 3.5% annual rate over the past 20 years. Recently, growth has slowed, and estimates for future growth vary widely. Published figures range from under 2% to almost 5% (8,9). While currently announced expansion plans imply that capacity will increase less than 3% per year between 1974 and 1977, such plans can change. Present US capacity is well under the amount needed to satisfy domestic demand, and it is entirely possible that the refining industry could expand at historic or higher rates in upcoming years even if demand growth slows. In the face of this great uncertainty, we chose to use 3.5% because it did not seem substantially worse than any other figure.

The CEP capital spending estimate is based on figures for refinery construction costs and the pollution control share of these costs (10,11). The estimate—described in detail in the Appendix—incorporates the following assumptions:

> 1. Pollution control capital needed to comply with 1977 air and water standards will add 10% to total refinery plant and equipment costs. Some 60% of this requirement has already been spent at existing capacity.
> 2. The industry will grow at a compound annual rate of 3.5%.
> 3. 1983 water pollution standards, which are significantly more demanding than those of 1977, will raise total pollution control capital needs to 12% of refinery construction costs.

The EPA has published several studies of air pollution control costs, one, "The Economics of Clean Air," as long ago as 1972. In June 1973 they presented a commissioned study by contractor Stephen Sobotka & Co. There is essentially only one EPA air estimate, however, since the Sobotka figures are based in large part on the earlier study. Both EPA estimates are founded partially on engineering studies and make some limited allowance for returns from air pollution control. On CEP's adjusted basis, EPA estimates indicate that $640 million in air pollution capital will be needed from 1974 to 1976. The EPA also presents estimates for the cost of using less polluting, low sulfur fuels. This is discussed below in the section on operating costs.

The EPA has also examined water pollution control costs. In 1972, the Agency commissioned contractor Roy F. Weston to study the costs of refinery water pollution control. The results of the report were released in the spring of 1973 and were reaffirmed in 1974 when the EPA published its final water control regulations (12). The report gives 1974-1977 capital requirements—on a CEP adjusted basis—equal to $1,000 million and estimates another $1,000 million will be needed to meet 1983 limitations. Equipment for both new and existing capacity is included.

In August 1973, the API published a Brown and Root, Inc. study of the future costs of refinery pollution control. The estimates were based on an engineering study that included surveys of current refinery control levels and on-site investigations. On CEP's adjusted basis, Brown and Root estimated that $1,150 million would be needed to meet control levels approximating 1977 water regulations and that an additional $1,925 million would be needed to reach 1983 control levels. An estimated $2,000 million more would be needed to eliminate pollutant discharges entirely.

Given the uncertainties of estimation and the wide margin for error, a range of figures, such as that in Table 5, offers the most realistic approach to estimating capital costs. A range is cumbersome to work with, however, so we have chosen to use a central figure of $3,500 million as the basis for further analysis.

CAPITAL RELATED PRODUCT COSTS

All of these capital estimates include both facilities to upgrade existing capacity and 100% of the pollution control capital needed to control new capacity. Such figures do provide a useful measure of how much money the industry will have to raise for pollution control, a presumably unprofitable enterprise, but capital estimates of this type can be misleading for almost any other purpose. People generally recognize that estimates for future capital requirements understate total control equipment since they do not include capital already in place. Less well recognized is that typical, all-inclusive future estimates, even after correction for revenue generation and cost savings, lead to an overstatement of added product costs if some pollution control has already taken place.

The equipment needed to bring new capacity to the level of control already achieved by existing refineries does not raise the average product cost because presently produced products already include these costs. Instead, this equipment will simply result in new refineries having pollution control costs equal to those already incorporated in the costs of products from existing refineries. Only that equipment which is needed to improve pollution control quality will lead to increased product cost. Only about 80% of the $3,500 million figure which we are using falls into this category.

Production costs rise not because equipment is installed, but because it costs money to own, operate and maintain. CEP has used a simple model to convert the capital estimates to an annual, cost-of-product basis. This model is described in detail in the Appendix, but its basic assumptions are:

    1. Only capital for improved levels of pollution control will increase unit product costs.
    2. Cost of net capital invested is 25% pretax.
    3. Depreciation is based on a 15 year life.
    4. Operating and maintenance expenses come to 10-20% of gross capital invested.
    5. The 7% investment tax credit is fully applicable.

The cost of capital in this model is essentially an opportunity cost element. It indicates the earnings foregone from not having the money invested in some profitable project. Because of this, the results are not directly analogous to income statement type costs. Rather, they are a measure of how much higher earnings might have been without these costs.

We cranked 80% of the total capital estimate through this model, and on this basis, calculated that capital related product costs would rise 11-14¢ per barrel pre-tax, 6-7¢ after tax. These results are independent of any assumptions about the industry's growth rate.

INCREASED OPERATING COSTS

Some pollution control can be achieved best by changing operating methods. Pollution control does not always require new or changed equipment. Sulfur oxides (SOx) in particular can be controlled in a variety of ways. SOx is formed when a sulfur-containing fuel is burned. A refinery burns sulfur-laden coke in the FCC regenerator and sulfur-containing process gases and fuel oils in furnaces and boilers. In some instances, SOx can be trapped and controlled after it is formed, but at present, the best control method is to burn fuels from which the sulfur has already been removed. Special equipment designed to desulfurize process gases and coke before burning adds to pollution control capital requirements. Other low sulfur fuels, low sulfur fuel oil and natural gas, for example, can be purchased. These cleaner fuels are generally more expensive than the alternatives, and the resulting additional costs represent increased operating costs that are quite closely related to the volume of refinery output.

CEP's cost estimates assume that some SOx emissions will be controlled through the use of low sulfur fuel oil. We assume that this fuel will supply 10% of refinery energy needs. We also assume that fuel consumption per barrel of product will not decline. At current refinery energy use rates, and assuming low sulfur fuels cost $1.20 per barrel more than the most likely alternative,* we calculate that using higher cost, low sulfur fuels will increase refinery costs by 1.5¢ per barrel of product. This is over and above the capital related costs estimated in the previous section. It is important to remember that more expensive low sulfur fuel oil raises operating costs and not capital requirements, even though the refinery may have installed equipment to desulfurize the fuel oil.

When refineries burn low sulfur fuel oils, they have become customers for their own clean burning products. As pointed out earlier, it is not appropriate to charge refinery pollution control with the costs of fuel oil desulfurization unless they use the fuels themselves. Process gas desulfurization, on the other hand, does constitute a proper refinery pollution control cost since this expense is undertaken to clean gases which will be used only within the refinery.

These operating costs could be reduced or replaced by capital related costs, especially if the margin of fuel provided by purchased, low sulfur fuels can be reduced. Process gases, produced as a by-product of refining operations, are a major refinery energy source. This in itself is a form of pollution control because these gases are generally flared off if they are not used for fuel. The amount of total energy the refiner gets from this source can be increased by redesigning refineries. The product configuration projected for Socal's El Segundo, California expansion, for example, indicates that this type of redesign may be taking place (14). Socal's Richmond, California refinery has undertaken a low sulfur fuels project which is expected to supply nonpolluting fuels to customers. It is

*In 1973, the EPA estimated that the long-run cost difference between high and low sulfur fuels will be 90¢ per barrel (13). The present difference is much higher because manufacturing capacity is limited, but the industry is expanding low sulfur fuel production facilities. The long-range difference thus seems appropriate for a ten year perspective. The 90¢ per barrel figure includes some underlying assumptions which differ from those used in this study. In particular, we use a higher cost of capital and higher operating and maintenance figures than the EPA does. Correction to a common assumption basis raises the EPA figure to $1.20 (1974 dollars) per barrel of fuel.

also expected to produce more by-product fuel gas than the refinery will need to fuel its new process units (15). Rising energy costs, combined with improvements in refinery technology, are making this an increasingly attractive alternative for new refinery construction although such refinery designs may be more expensive than conventional designs. Process gases, however, frequently have a high sulfur content, and equipment must be installed to remove the sulfur before they are burned. Net pollution control costs may or may not be lower.

Another way to reduce the pollution fuel burning causes is to reduce net energy consumption. The industry's energy use per unit of product has been falling since 1946 as processing efficiency has improved. Lower air pollution levels are a welcome by-product of this higher efficiency, and given air pollution regulations, improve the return from energy use conservation. Both energy costs and pollution control charges are saved. Conserving energy has its limits as an air pollution measure, however, and the problem of polluting fuels remains. For the immediate future, SOx emissions are most likely to be controlled by desulfurizing fuel before it is burned.

Water pollution control costs can also be affected by operating changes. The amount of water pollution that flows from a refinery is directly related to the quality of its housekeeping practices. According to the EPA, good housekeeping includes "minimizing waste when sampling product lines; using vacuum trucks or dry cleaning methods to clean up oil spills; using a good maintenance program to keep the refinery as leakproof as possible; and individually treating waste streams with special characteristics . . . . (16)." Extra dollars spent on refinery upkeep, maintenance, and general cleanliness can save on pollution control down the line. It can lower capital investment requirements as the size of the treatment unit needed is reduced, and it can lower operating and maintenance costs for a given unit. Exxon found that carefully monitoring leaks and spills at one refinery saved as much as $3,000 a day (17).

CEP does not have figures for the added costs of improved housekeeping, and we have made no attempt to quantify the amount of expenditure involved. This omission probably means our net estimated control costs are too high because higher housekeeping costs generally are more than offset by other reduced costs of operations, including pollution control costs.

Improved housekeeping is not the only way a refiner can trade higher operating costs for reduced capital investment. The refinery sometimes has the option of paying someone else to treat its polluted wastewater rather than treating it on-site. Some refiners are near municipal or other joint treatment plants and can discharge to them. This does not absolve the refiner of costs. A municipal or other joint plant generally charges for treatment. The price is commonly based on such features as the volume of water treated and how polluted it is. The price is supposed to cover full treatment costs. In addition, the refinery may have to provide primary treatment before the municipal plant will take the discharge. The refiner thus avoids the capital expenditure to install secondary and tertiary treatment and gets increased operating costs for these phases of control. Economies of scale in treatment suggest that the municipal or joint treatment route may be cheaper overall. Recent EPA action may mean, however, that municipal plants will be subject to more severe restrictions than they have enjoyed in the past, and that may mean that refineries which discharge to them will have to do a more thorough primary treatment job.

The joint treatment option is not available to every refinery, but when it is, the refiner often chooses to use it. Eight of the 61 refineries covered in *Cracking Down* discharge their wastewater to municipal treatment plants. CEP's cost estimates generally assume that all refineries will be treating their own wastewater. We make no adjustments to capital investment or operating cost estimates to allow for the possible use of these municipal treatment facilities. The primary implication is that if refineries use municipal treatment to any great extent, our capital cost estimates may be too high and operating cost estimates too low.

RETROFITTING VS. DE NOVO CONTROL

New and better design and the ability to make control equipment an integral part of a refinery often mean that new capacity will have lower control costs per unit of product. Such savings are not realized, however, unless they are built in at the beginning. If environmental regulation appears likely, there should be a strong motivation for a company to think ahead at the time of construction rather than having to retrofit later. As one Socal executive put it to CEP, "Staying two steps ahead of the sheriff can make good business sense."

Expansions at old refineries may have different costs than entirely new construction will have. Without knowing something about the refinery to be expanded, no one can predict which will have the cheaper cost. An old refinery might have unused control capacity which the expansion could use. In addition, pollution control techniques are often unique to a particular refinery design, and for the refinery which has already learned by doing, the costs may be less than for an entirely new unit. The new refinery, on the other hand, will have the advantage of not having to work around old equipment. No matter which is cheaper, the difference in costs for the new capacity versus those for retrofitting can be substantial.

New refineries should cost less to control than old, but the reduction in average costs for the refining industry as a whole will not be substantial for many years. Even at a 3.5% per year annual growth rate, new post-1974 capacity will make up only 26% of the industry's total refining ability by 1983. CEP calculates that if new control is 10-20% cheaper than retrofitting, average pollution control costs for the refining industry as a whole will fall by 3-5% in 1983. All our calculations are based on costs at existing refineries and make no attempt to allow for cheaper de novo control. This is one reason that assumed expansion rates do not affect the calculations. To the extent that de novo control is cheaper, however, estimated costs for the industry, and particularly the individual new refinery, will be overstated.

AN INFLATED ESTIMATE

Refining pollution control is costly. Our estimates show added costs averaging as high as 11-14¢ per barrel between 1974 and 1983. Total refining costs for turning crude oil into products come to only $1.50-2.00 per barrel, so the relative size of pollution control costs is large. The pollution control cost estimates presented in this study are likely to overstate the true costs; some of the reasons for this overstatement have been suggested in Chapter 3. It is simply impossible to account adequately for the possible returns from control. Those returns may be significant and are very likely to increase over time as technology finds new ways to use pollution control by-products.

Two further sources of overstatement in the estimates are analytical. Some may consider the cost of capital used to be generous, and no allowance is made for cheaper costs of control at new refineries. Neither factor is likely to be significant in the immediate future, but cheaper de novo control, in particular, may well be a factor for the long run.

Additional difficulties arise because all estimates that are made today must be based on equipment available today, or at least on the drawing board. It is hard to believe, however, that the highly innovative petroleum industry will not improve its control techniques in the next few years, especially given the cost and regulatory pressures that are being put upon them. New methods, if implemented, should be cheaper than old or they would not be employed. For this reason alone, we should expect that future control costs (always in constant dollars) will be less than those implied by present methodology.

REFERENCES

1. "Oil Industry Spending Billions to Comply with Requirements for Unleaded Gasoline," *Wall Street Journal*, 29 June 1974; Robert O. Skamser, "U.S. Refining Must Double Its Annual Investment for 1980's Needs," *Oil and Gas Journal*, 27 August 1973; and Stephen Sobotka & Company, "Economic Analysis of Proposed Effluent Guidelines, Petroleum Refining Industry, Part I," prepared for the Environmental Protection Agency, Office of Planning and Evaluation, September 1973, Exhibit 16.
2. Environmental Protection Agency, "The Economics of Clean Air," Annual Report of the Administrator of the Environmental Protection Agency to the Congress of the United States, March 1972.
3. Stephen Sobotka & Company, "The Impact of Costs Associated with New Environmental Standards upon the Petroleum Refining Industry," prepared for the Council on Environmental Quality, November 23, 1971; and "Economic Analysis of Proposed Effluent Guidelines, Petroleum Refining Industry," Part I prepared by Stephen Sobotka & Company, Part II prepared by the EPA, September 1973.

4. Roy F. Weston, "Development Document for Proposed Effluent Limitation Guidelines and New Source Performance Standards for the Petroleum Refining Industry," prepared for the Environmental Protection Agency, EPA 440/1 - 73/014, Point Source Category, December 1973.
5. Brown & Root, Inc., "Economics of Refinery Wastewater Treatment," prepared for the American Petroleum Institute, Committee on Economic Affairs, API Publication #4199, August 1973.
6. American Petroleum Institute, "Environmental Expenditures of the US Petroleum Industry, 1966-1972," API Publication #4176, 1973.
7. John E. Cremens, "Capital Expenditures by Business for Air and Water Pollution Abatement, 1973 and Planned 1974," *Survey of Current Business*, July 1974, pp. 58-64.
8. Sobotka, 1973; and "Refining Capacity in US to Jump 13% by 1978," *Oil and Gas Journal,* 20 May 1974, p. 56.
9. Brown and Root, Inc.; and Skamser.
10. E. K. Grigsby, E. W. Mills, and D. C. Collins, "Capital Requirements for Refined Petroleum Products," paper presented at the National Petroleum Refiners Association Annual Meeting, 1-3 April 1973, San Antonio, Texas.
11. W. L. Nelson, "Questions on the Technology, Refinery Pollution Control," *Oil and Gas Journal,* 4 September 1972, p. 86; and W. L. Nelson, "Costs of Refineries, Part 4," *Oil and Gas Journal,* 29 July 1974, p. 162.
12. Environmental Protection Agency, "Effluent Guidelines, Petroleum Refining Industry," *Federal Register,* XXXIX, No. 91, Part II (9 May 1974) pp. 16574-16575.
13. Sobotka, 1971, pp. 3, 14-16.
14. Gregg Kerlin and Daniel Rabovsky, *Cracking Down: Oil Refining and Pollution Control* (New York: Council on Economic Priorities, 1975), refinery profile, Standard Oil, El Segundo, California
15. Kerlin and Rabovsky, refinery profile, Standard Oil, Richmond, California.
16. Weston, pages 95-6.
17. Kerlin and Rabovsky, Chapter 3.

5: The Big Eight Refiners

Eight large companies dominate the petroleum refining industry in the United States. In the companion to this volume, *Cracking Down*, CEP assesses the quality of pollution control each of the big eight refiners has achieved. In this chapter, we apply the results of that pollution control quality study to the costs of individual firms. In particular, we have estimated how much full compliance with government pollution control regulations will affect these company costs and profits over the next ten years.

All of the eight largest refiners are vertically integrated petroleum companies. Together, they own approximately 55% of US refining capacity. The largest refiner, Exxon, controls almost 9% of the total, Arco, the smallest of the big eight, accounts for 5.5%. Figure 9 illustrates the relative capacities of the 20 largest US refiners and the clear break between Arco and the ninth largest firm, Sun Oil (see Table 4).

Cracking Down examined air and water pollution control at each of the 61 refineries and found that control ranges from very poor to quite good. In addi-

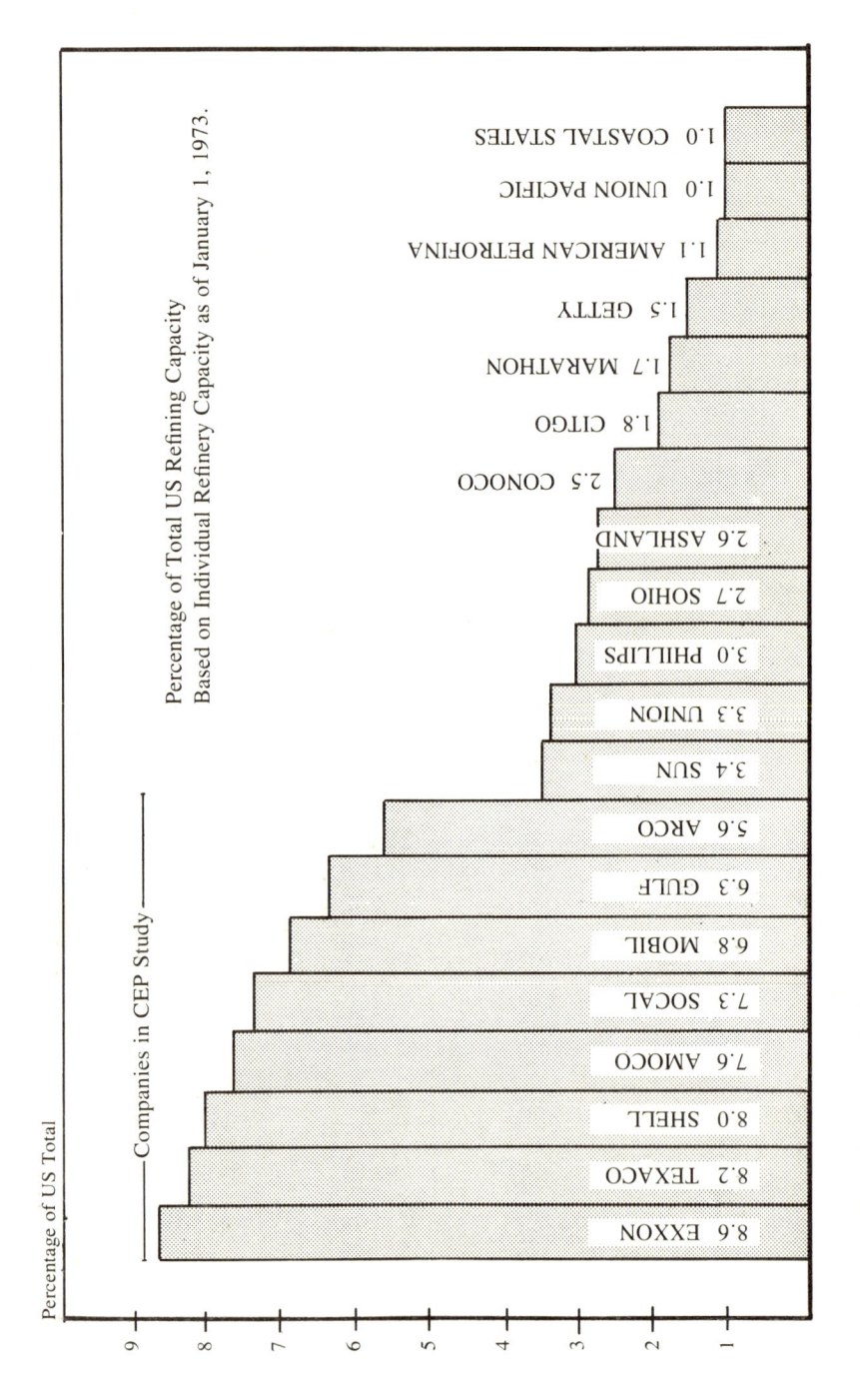

FIGURE 9. The 20 Largest Refiners

Percentage of Total US Refining Capacity Based on Individual Refinery Capacity as of January 1, 1973.

Company	Percentage of US Total
EXXON	8.6
TEXACO	8.2
SHELL	8.0
AMOCO	7.6
SOCAL	7.3
MOBIL	6.8
GULF	6.3
ARCO	5.6
SUN	3.4
UNION	3.3
PHILLIPS	3.0
SOHIO	2.7
ASHLAND	2.6
CONOCO	2.5
CITGO	1.8
MARATHON	1.7
GETTY	1.5
AMERICAN PETROFINA	1.1
UNION PACIFIC	1.0
COASTAL STATES	1.0

Companies in CEP Study: ARCO through EXXON (top 8)

TABLE 6

The Big Eight

Company	U.S. Capacity (bbl/d)	Number of U.S.* Refineries	Average Capacity (bbl/d)
Exxon	1,252,000	5	250,000
Shell	1,127,000	8	141,000
Amoco	1,043,000	8	130,000
Texaco	1,037,000	11	94,000
Socal	952,000	9	106,000
Mobil	887,000	7	127,000
Gulf	860,000	8	108,000
Arco	785,000	5	157,000
BIG EIGHT TOTAL	7,943,000	61	130,000
INDUSTRY TOTAL	14,216,000	247	58,000

*Only finished fuels refineries for Big Eight companies.

tion, CEP found that a number of refineries with good water pollution control records are major air polluters, and vice versa. The differences among overall company performances are somewhat less dramatic, but there is a discernible range. Figure 10 illustrates the relative positions of the firms from best to worst for both air and water pollution control.*

Full compliance with pollution control regulations will affect company costs and profits to varying degrees in the future. This is despite the fact that we assume that if all eight companies were starting from scratch, they would each have to spend the same amount, roughly $250 per barrel of capacity, to install the necessary pollution control capital. The companies are not starting from scratch. The costs and profits of today, to which we wish to compare future performance, reflect pollution control investments that were made in the past. Companies that are doing a poor control job now have effectively inflated their profits by postponing control costs. Well controlled firms should have depressed profits compared to their competitors. Prospects for the future, compared to today's situation, depend on how much or how little of the pollution control job is left to be done.

*See Chapter 5 in *Cracking Down* for the rating system used to compare refinery performance. This system compares refineries to one another rather than to some absolute standard.

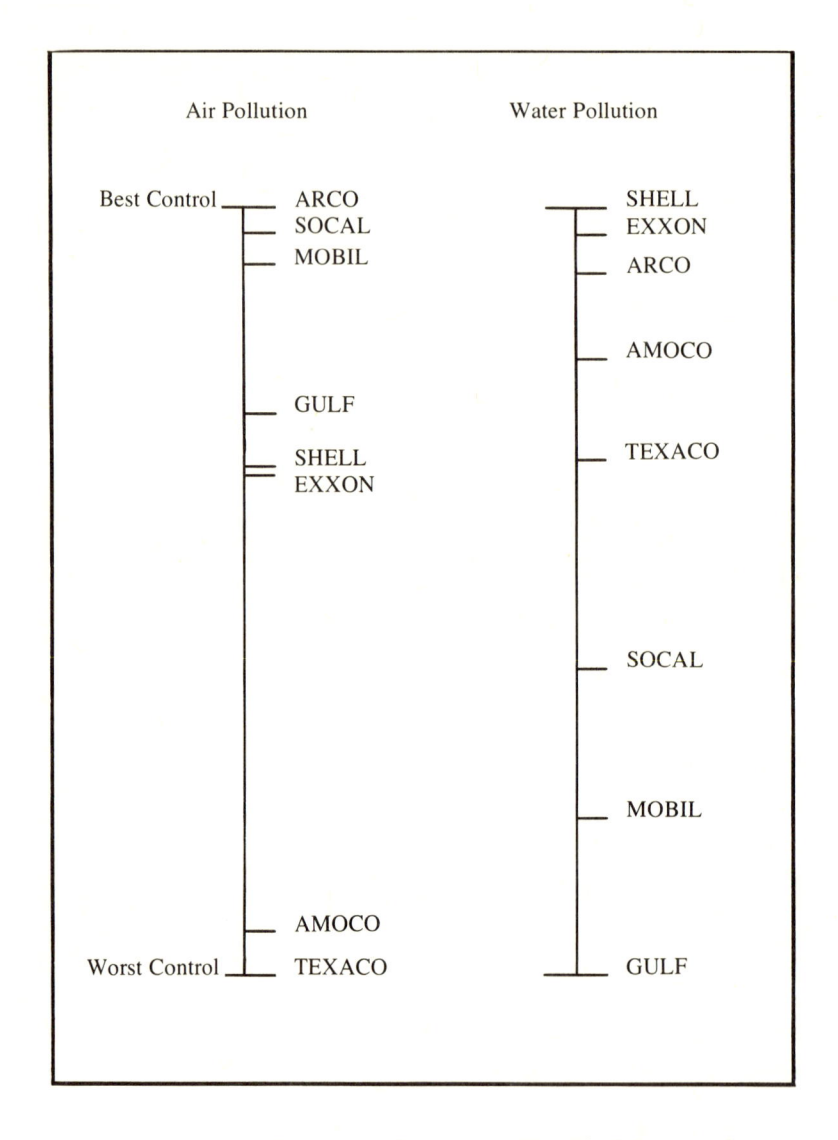

```
                Air Pollution              Water Pollution

Best Control ──┬── ARCO              ──┬── SHELL
               ├── SOCAL              ├── EXXON
               ├── MOBIL              ├── ARCO

                                       ├── AMOCO

               ├── GULF
                                       ├── TEXACO
               ├── SHELL
               ├── EXXON




                                       ├── SOCAL



                                       ├── MOBIL



               ├── AMOCO
Worst Control ─┴── TEXACO             ─┴── GULF
```

FIGURE 10. Pollution Control Performance—The Big 8 Ranked from Best to Worst

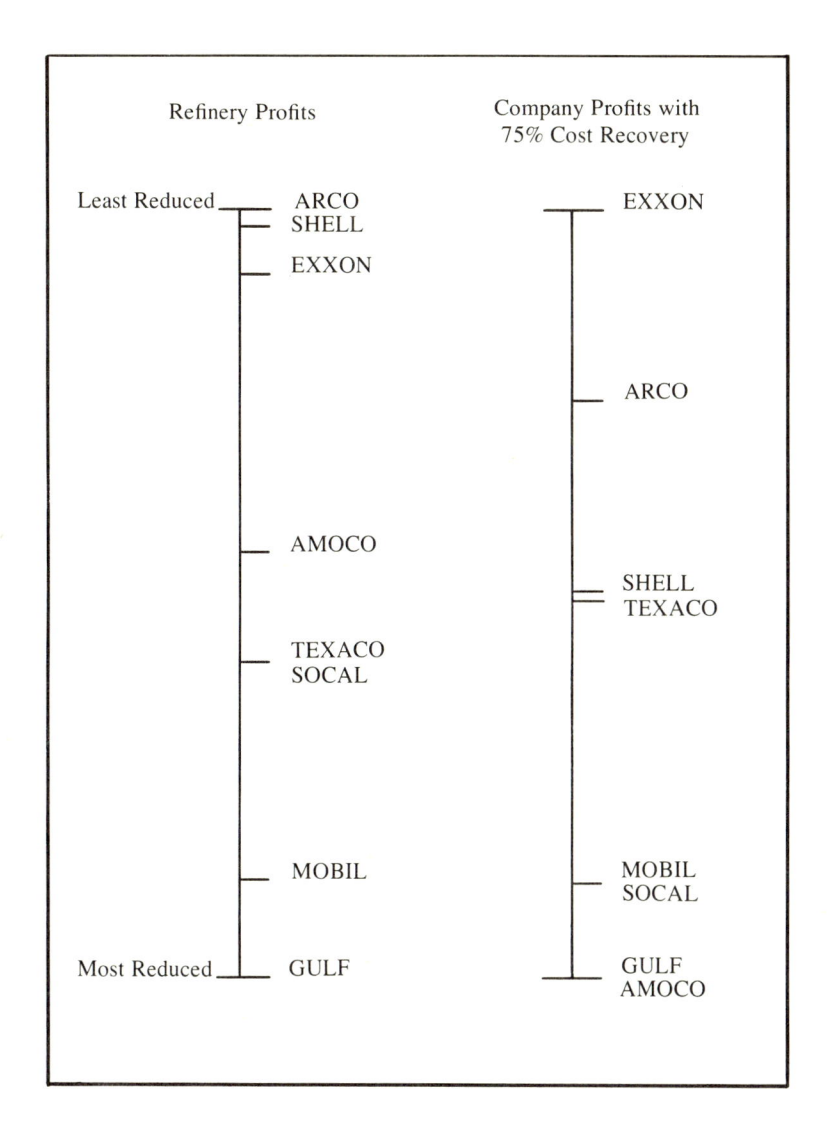

FIGURE 11. Pollution Control Cost Impact on Big 8 Profits

We assume that differences in the quality of present control mean that in the past, firms have invested different amounts in pollution control and that they have different proportions of the $250 per barrel total left to spend. The assessments of current control performance are not defined in terms of an absolute standard which the refineries must meet, but in terms of company by company comparison. Given this analytical foundation, individual company cost figures have less significance standing alone than they do in relation to one another. Figure 11 illustrates the results in brief. The company profiles at the end of this chapter discuss the impact future pollution control costs will have on individual companies.

ESTIMATING COMPANY COSTS

The pollution control capital requirement estimates presented in Chapters 2 and 4 are all industry-wide. It was necessary to devise some way of allocating these industry totals to the companies in order to estimate costs for individual firms. CEP has used the percentage of total US refinery capacity operated by an individual firm—with rough adjustments to reflect differences in quality of control and differences in planned expansions—as the base for allocating capital equipment requirements. We assume that all firms will face an additional charge of 1.5¢ per product barrel for more costly, cleaner burning fuels.

We assume that the average big eight firm will have to spend the same amount of money between 1974 and 1983, per barrel of capacity, as the average refiner in the industry. In particular, we assume the industry has already spent 60% of the monies necessary to comply with air standards and 1977 water standards. We assume that all of the additional equipment necessary to improve water control from 1977 to 1983 required levels is still to be installed.

It is necessary to digress to point out that it is not clear whether the big eight firms, on the average, lead or lag behind the rest of the industry. They may be technologically more advanced and therefore ahead of smaller firms. On the other hand, some of their competitors's refineries may be smaller, simpler, and less costly to control. In addition, the large refiners so dominate statistics on the industry that if they are performing significantly above the industry average, other firms must be correspondingly below. This is a matter of simple arithmetic. If the industry has spent an average of 60% of its required capital amount, and the big eight has spent 65%, this means the refineries operated by other firms will have invested only 54% of their needed capital amounts. Rather than make an assumption for which we have little factual basis, we simply go along with the industry average.

It should also be pointed out that 60% of capital in place implies that a good deal more than 60% of the pollution control job is done. The last units of pollution are by far the most expensive and difficult to control. The remaining 40% of capital investment and costs may be required, for example, to abate the last 5-10% of pollutants.

CEP calculated the new pollution control costs per barrel of product for each firm on the basis of costs to bring existing 1974 capacity into compliance with pollution control regulations. In addition, we estimated the amount of pollution control investment which will be needed to control any expanded capacity which the firm adds between 1974 and 1983. The total pollution control capital figure for each firm thus has two parts: 1. capital to complete control of existing refineries. This is the investment which will indicate how company production costs will rise from current levels. 2. Capital needed to make any new capacity environmentally acceptable. Together, these two capital elements indicate the 1974-1983 pollution control capital burden on the company, a financial burden which is presumably unprofitable (but see Chapter 3 for a discussion of this matter).

Our estimates for pollution control capital needed to bring existing capacity up to 1977 standards depends on present control levels. We assume that additional investment to complete the water control job necessary to meet 1983 requirements will be the same for each firm per barrel of capacity. Here, the allocation basis is simply a company's percentage of industry capacity.

CEP constructed relative cost indexes for each firm to translate control performance into costs. These indexes, which show which firm has the largest prospective costs per capacity barrel, have a capacity weighted average of 1.00. This reflects our assumption that the big eight average capital installation equals the industry average. Table 7 presents our calculated index along with each firm's average emissions for selected air and water pollutants. These are the figures on which the index is based. The index number is used as follows:

Company Capital Allocation Factor =
(Cost Index) (Company % of Industry Capacity)

Company Capital Requirement =
(Company Allocation Factor) (Industry Capital Estimate)

The relative cost index comes from comparing a firm's 1974 actual control performance to that of its peers. We treat air and water pollution separately because we assume a company's air control problem is less expensive to deal with than water pollution control. Water pollution performance gets approximately 3 times the weight air of air quality in the index construction.

Water pollution control is a complex matter, and a number of pollutants must be considered in any evaluation of company performance. Some of these pollutants are more costly to control than others. We have tried to make a rough allowance for this and for any potential returns to control. For example, because controlling oil and grease pollution brings some return to the polluter and relies on a fairly well developed control technology, we rate it as only half as costly to control as BOD and COD. A company's relative quality of control for each major water pollutant is combined with the assumed difficulty factor to achieve an overall water pollution control cost index. This gives a measure of the level of cost for a company compared to the other seven firms. In some cases, these

TABLE 7

Company Effluent and Emissions Levels with CEP Calculated Cost Indexes*

	Arco	Exxon	Gulf	Mobil	Shell	Socal	Amoco	Texaco
Air Pollution Emissions: Tons /1000 bbl/d								
Particulates	.015	.025	.024	.012	.020	.014	.027	.017
SOx	.162	.150	.167	.210	.289	.188	.325	.338
Hydrocarbons	.104	.219	.186	.143	.196	.116	.220	.350
CO	.01	.173	.20	—	—	.056	1.179	1.622
Cost Index	.56	.94	.86	.65	.94	.61	1.59	1.66
Water Pollution Effluent: Net pounds per 1000 bbl/d								
BOD	9.5	22.2	53.3	45.4	4.1	34.0	17.7	28.7
COD	87.2	40.7	183.3	100.9	56.6	134.6	105.9	58.6
Oil and Grease	5.3	4.2	30.5	33.7	1.8	8.3	10.9	13.4
Ammonia	2.6	11.6	5.2	18.7	9.2	29.4	6.4	1.9
Phenol	2.6	(0.1)	6.3	2.7	1.8	1.7	1.2	2.4
Cost Index	.62	.55	1.93	1.63	.49	1.36	.78	.98
Combined Cost Index:								
Raw	.60	.65	1.65	1.40	.60	1.15	1.00	1.15
Adjusted	.73	.77	1.43	1.27	.73	1.10	1.00	1.10

*An index of this sort is a ratio derived from a series of observations and used as an indicator or measure to express the ratio of one dimension of a thing to another dimension. In this case, the index is a way of saying that one company's costs are X% of the average industry costs.

indexes were then adjusted to reflect special information that was gained in the gathering of data for *Cracking Down*. If, for example, a given company's result is heavily distorted by a single refinery, this might be taken into account.

The whole process is repeated for air pollution. Here, pollutants such as CO and hydrocarbons, which can bring substantial returns to the polluter when they are controlled, have very little weight in the calculation. Primary cost emphasis is put on SOx and particulate problems.

We then combined each company's air and water index into a total cost index. Water control performance was given much the greater weight in the combined index, not only because water pollution is more expensive to control, but because most of the future clean-up costs are water connected. We made one additional adjustment to the combined index before calling it final. Each calculated value was reduced by one-third its distance from the eight company mean. Thus, a raw value of .70 becomes .80 for calculation. Similarly, 1.30 becomes 1.20.

Equipment needs for expansion are based on a number of assumptions. To begin, we assume that the industry will grow at an average annual rate of 3.5% a year. Individual company expansion assumptions are based on company an-

TABLE 8

Company Shares of Total Refining Industry Capacity and Expansion With Capital Allocation Factors.

	1974 Share of Industry Capacity	Capital Allocation Factors		
		Share of '74 Capacity	Share of '74-'77 Expansion	Share of '78-'83 Industry Spending
Arco	5.5%	4.8%	5.5%	5.5%
Exxon	8.8%	5.9%	10.5%	9.0%
Gulf*	6.0%	7.0%	1.7%	5.5%
Mobil	6.2%	7.6%	1.7%	5.7%
Shell	7.9%	6.3%	2.2%	7.2%
Socal	6.7%	6.9%	20.0%	8.5%
Amoco*	7.3%	9.3%	2.0%	6.7%
Texaco	7.3%	8.0%	3.0%	6.7%
Rest of Industry	44.3%	44.2%	53.9%	45.2%
TOTAL	100.0%	100.0%	100.0%	100.0%

*Both Gulf and Amoco have announced tentative expansion plans greater than those indicated here. These plans have not been confirmed, and construction appears to be a long way off. In addition, Texaco has postponed some major expansions indefinitely (3). None of these tentative or postponed expansions is included.

nouncements about construction that is under way or at least in the design and engineering stage (2). We have ascribed a minimum growth rate of one per cent a year even to those companies with no announced expansion plans. A firm can achieve such a rate even without new construction simply by relieving bottlenecks at existing refineries. Using these company figures and the assumed industry growth rate, we calculated the proportion of industry expansion assignable to a given firm. This takes care of the expansion estimate through 1977. After 1977 we assume that both the industry and the individual firms will grow at a common 3.5% a year. Table 8 indicates the individual company shares of capacity (also illustrated in Figure 9) and expansion through 1977 as well as the allocation figures CEP used to estimate the capital that will be needed to upgrade existing refineries.

The individual company capital estimates which result from the allocation are summarized in Table 9 along with figures indicating the total capital amounts which will go to improve average company control levels. This latter amount of the total is the only portion which will affect costs per barrel.

TABLE 9

Capital Investment Requirements for Pollution Control

(All figures in millions of 1974 dollars.)

Company	1974-1977 Upgrade Existing Capacity	1974-1977 Conform To New Capacity	1977-83 Total Investment	1974-83 Total Investment	% Total Needed to Improve Average Level of Pollution Control
Arco	55	24.0	90	169	67%
Exxon	91	44.0	143	278	68%
Gulf	116	7.5	87	210	80%
Mobil	107	7.5	93	208	77%
Shell	78	9.7	114	202	67%
Socal	99	88.0	135	322	75%
Amoco	99	8.8	106	214	73%
Texaco	108	13.0	106	227	75%
Rest of Industry	647	247.0	776	1670	73%
TOTAL	1400	450.0	1650	3500	73%

Still to Spend

1974 Dollars

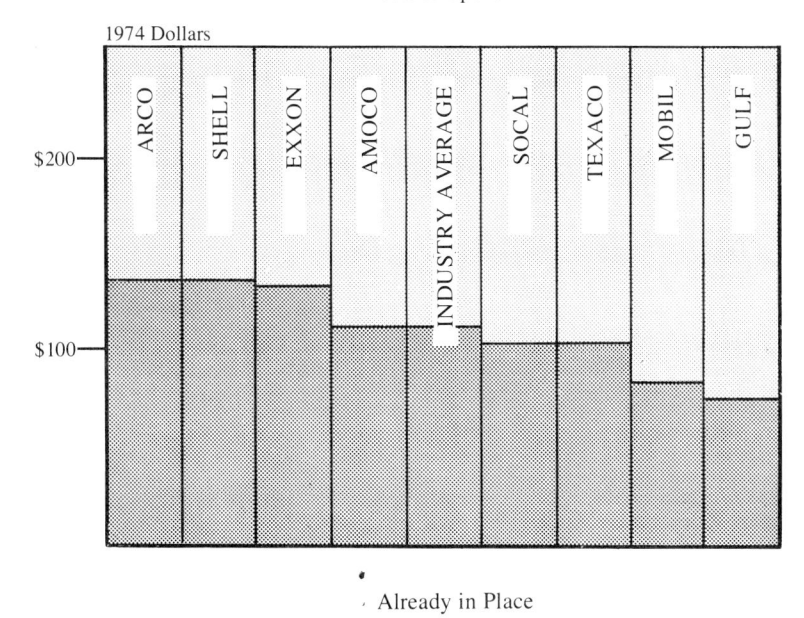

[Bar chart showing ARCO, SHELL, EXXON, AMOCO, INDUSTRY AVERAGE, SOCAL, TEXACO, MOBIL, GULF with $200 and $100 reference lines]

Already in Place

FIGURE 12. Capital Requirements to Control Pollution per Barrel of Capacity

Figure 12 shows the CEP calculated breakdown between investment already made for pollution control versus capital still required for each of the firms on a per barrel basis. This assumes that each firm will have to spend an average total of $250 per barrel for control.

CEP used the model described for the industry cost estimates (see Chapter 4 and the Appendix) to convert company capital estimates to a product cost basis. Once again, no allowance is made for possible cheaper costs of controlling new capacity. This means that assumptions about expansion rates will not affect the results, thereby removing a possible large source of error, but it also means that costs for companies with especially high expansion rates will be overstated compared to costs for firms which are standing pat. Socal, for example, plans to expand more than any other company. We estimate that by 1983 over 40% of Socal's total capacity will have been added since 1974. Twenty-six per cent of the industry's 1983 capacity will have been built since 1974. The slowest growing companies—Mobil, Shell, Amoco and Gulf—will have, by our figures, 19% new capacity. We consider the possible cost significance of expansion in the company profiles.

ESTIMATING COMPANY COSTS 61

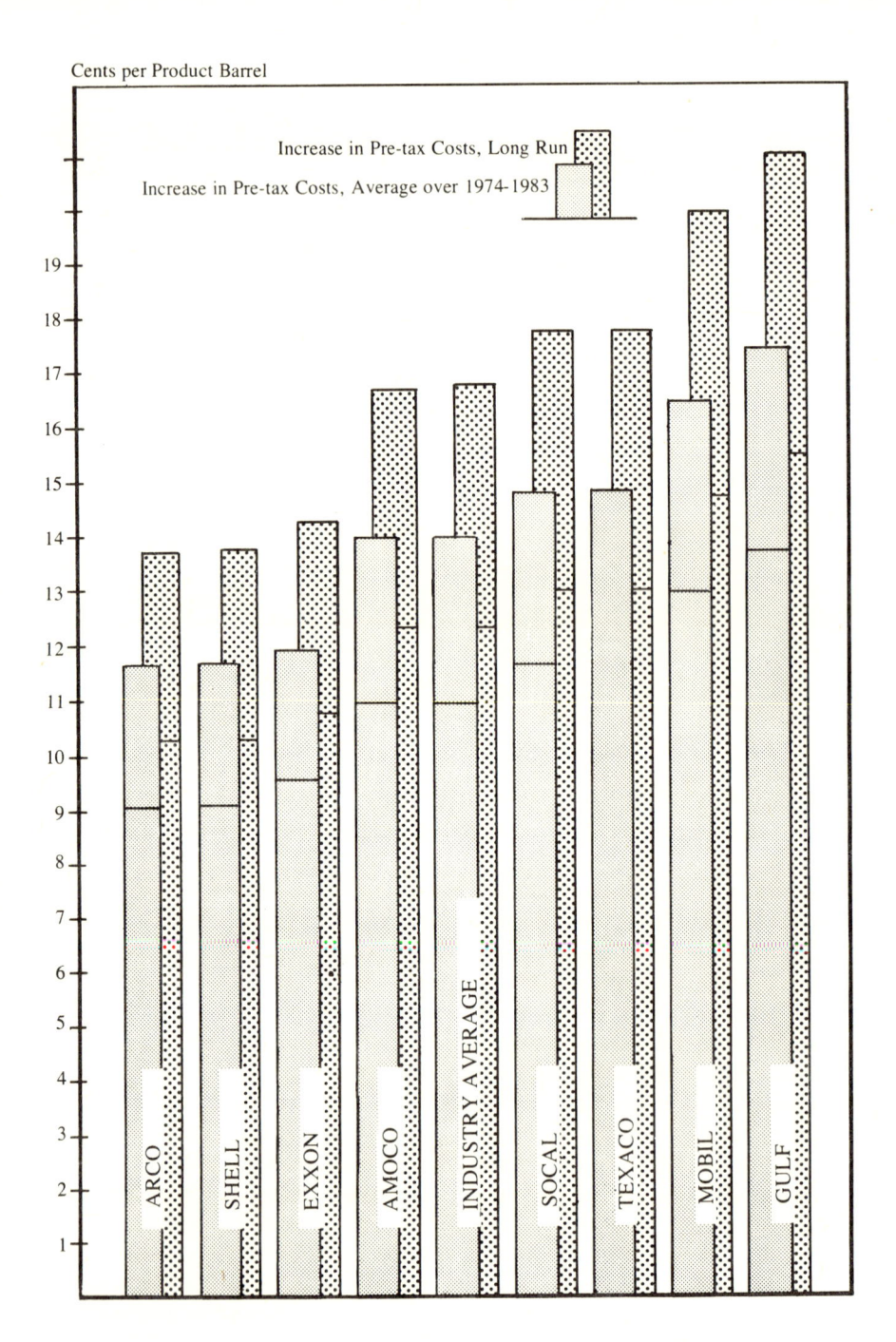

FIGURE 13. The Costs that Meeting 1983 Federal Standards Add per Barrel

62

Figure 13 shows CEP's estimates for the cost per product barrel for each firm to come into complete compliance. The light bars indicate the average costs over the decade 1974-83. The dark bars show the long-run costs of moving from today's control levels to mandated standards.

COMPARING COSTS TO PROFITS

We have not attempted to make careful profit or investment projections. We have merely assumed that company investment levels and profits will grow at the same real rate as we assume for the economy over the next ten years, 4% annually. Slower or faster growth rates for individual companies will alter our results in a corresponding fashion. One simple solution would have been to assume that profits and company investment would grow at the same rate as refining capacity. That overemphasizes the importance of refining to a petroleum company's total operations, however, and we have no reason to assume that other parts of the company will grow at the same rate.

Profit impact figures are useful mainly to compare one company to another, one industry to another. There are some problems in comparing our figures for petroleum company profit impacts to companies in other industries. Certain standard accounting practices in the petroleum industry, especially with respect to capitalization of investments, differ quite substantially from those of other manufacturing industries. Comparing rates of return on investment or net worth between this industry and others is very tricky. Petroleum profits may be understated relative to those of other industries, but even this is uncertain. The opposite could be true.* It is also difficult to make comparisons within the industry. Texaco's accounting methods, for example, differ from those of its competitors (4). Although CEP compares pollution control costs to published profit figures which may not be strictly comparable, the results of our calculations are intended only to indicate orders of magnitude. Under these circumstances, differences in accounting niceties do not seem to be a real problem.

The size of company profit reduction depends on the amount of each firm's incremental costs and on how much of that cost is recovered through product price increases. Industry-wide prices are the norm in the petroleum business, and no one firm can maintain a price significantly different from its competitors for any extended period of time. For this analysis, then, we must think in terms of industry average cost recovery. In the past, industry-wide price increases which have covered average industry pollution control costs have actually improved the profits of ill controlled refiners, assuming they have spent less money on control.

---

*Reported rates of return are intimately related to accounting practices. The petroleum industry tends to expense some expenditures which other businesses capitalize, resulting in a reduced book value of net assets. Reported profits and taxes, however, are also reduced by this type of accounting. With both the numerator, income, and the denominator, net assets or equity, reduced, it is hard to say which way the rate of return will go.

At the same time they have failed to preserve fully the profits of well controlled firms.

The shoe is on the other foot for the future. The well controlled refiner of 1974 should be able to recover future costs of control and perhaps a bit more via industry-wide price increases. Even full industry cost recovery will not maintain the profits of the firm which has lagged and has a great deal of spending to do to catch up. Figure 14 shows CEP's estimate of profit reductions for each firm, first if there is no cost recovery, and then assuming that, on the average, the industry manages to recover 75% of pollution control costs from the public.

Refining cost increases and company profit reductions are not perfectly correlated (see Figure 12). The importance of domestic refining in these companies' operations varies from just under 20% of corporate assets for Arco to around 4% for Exxon (5,6,7,8,9,10,11,12). This difference in emphasis on US refining, more than the differences in projected refinery costs per barrel, accounts for the range of estimated profit impacts.

Nonetheless, separately stated refining profit impacts are not meaningless. The long range evaluation of multinational oil company profits must take these into account since, as other countries impose pollution control regulations similar to those in the US, refinery operations abroad may encounter profit reductions akin to those experienced in the United States. Exxon, Gulf, and Texaco, with the largest proportion of nondomestic refining capacity, will face additional potential declines in company returns that domestic refiners such as Shell* and Arco will not experience.

Table 10 summarizes some of the results illustrated in Figures 12 and 13 in this chapter. More detail will be found in each of the company profiles. We have analyzed the first company, Arco, in most detail. Much of what is said in the Arco profile applies to the other companies as well.

*We are dealing here only with the US arm of Royal Dutch Shell, itself a very large multinational firm. Shell US is a semi-independent company with publicly traded stock which the general public can buy.

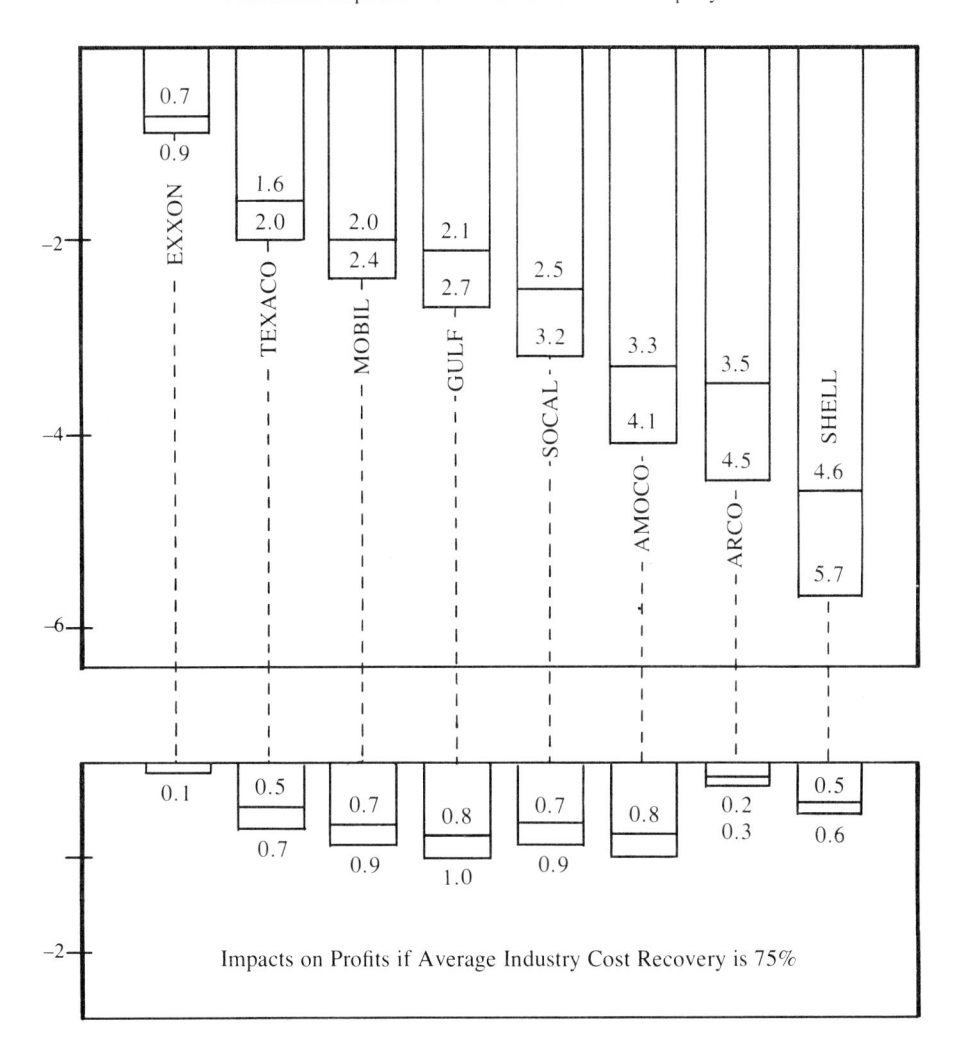

Maximum Impact on Cumulative 1974-83 Company Profits

EXXON	TEXACO	MOBIL	GULF	SOCAL	AMOCO	ARCO	SHELL
0.7 / 0.9	1.6 / 2.0	2.0 / 2.4	2.1 / 2.7	2.5 / 3.2	3.3 / 4.1	3.5 / 4.5	4.6 / 5.7
0.1	0.5 / 0.7	0.7 / 0.9	0.8 / 1.0	0.7 / 0.9	0.8	0.2 / 0.3	0.5 / 0.6

Impacts on Profits if Average Industry Cost Recovery is 75%

FIGURE 14. Effect on Profits from Completing Compliance with Pollution Control Regulations

TABLE 10

Summary of Results

(All values are in 1974 dollars.)

Company	1974-1983 Capital Investment for Pollution Control — Total	Existing Capacity (cost/bbl)	Post 1973 Product Cost Increase* ¢/product bbl — 1974-83 Average	Long Run Average	% Reduction in 1974-83 Inclusive Company Profits — Maximum: Zero Cost Recovery	Probable: 75% Industry Average Cost Recovery
Arco	$169,000,000	$111	9.0–11.4	10.0–13.5	3.5%–4.5%	.2%–.3%
Exxon	$278,000,000	$116	9.3–11.7	10.4–14.0	.7%–.9%	.1%
Gulf	$210,000,000	$178	13.5–17.2	15.2–20.7	2.1%–2.7%	.8%–1.1%
Mobil	$208,000,000	$167	12.8–16.3	14.4–19.6	2.0%–2.4%	.7%–.9%
Shell	$202,000,000	$112	9.0–11.5	10.1–13.6	4.6%–5.7%	.5%–.6%
Socal	$322,000,000	$147	11.5–14.6	12.8–17.4	2.5%–3.2%	.7%–.9%
Amoco	$214,000,000	$138	10.8–13.7	12.1–16.4	3.3%–4.1%	.8%–1.0%
Texaco	$227,000,000	$147	11.5–14.6	12.8–17.4	1.6%–2.0%	.5%–.7%
INDUSTRY	$3,500,000,000	$138	10.8–13.7	12.1–16.4		

*Assumes 7% investment tax credit.

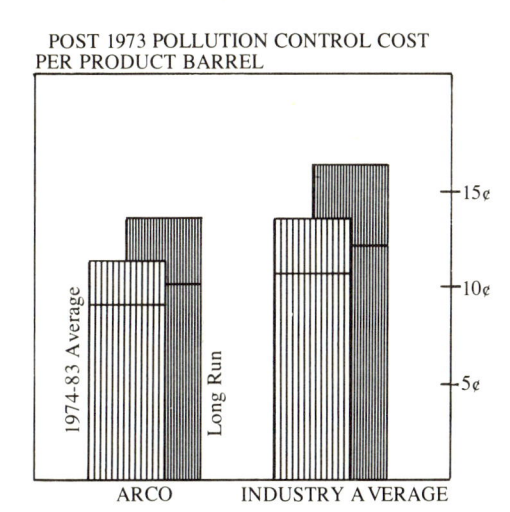

ATLANTIC RICHFIELD CORPORATION

POST 1973 POLLUTION CONTROL COST PER PRODUCT BARREL

[Bar chart showing ARCO and INDUSTRY AVERAGE, with bars for 1974-83 Average and Long Run, scaled in cents (5¢, 10¢, 15¢)]

Atlantic Richfield Corporation (Arco) is the smallest of the eight petroleum companies in this study, and it is also the smallest refiner. The firm is small, however, only in relation to its oil company peers. Arco was the 17th largest industrial in the United States in 1973, measured by assets, 26th in terms of sales, and 24th by profits (13). Profits in 1974 came to $475 million (14), more than the *revenues* of all but the very largest corporations. The company's return on equity, however, does not look as good. The 1973 rate of 8.9% on the stockholder dollar was dead last among the eight oil companies and 400th on *Fortune's* list of 500 industrials.

Arco is largely a domestic oil company rather than a multinational firm. US refining is important to the company's operations; it accounts for almost 20% of total corporate investment (5). Only Shell has as large a stake in domestic refining. Thus, while Arco is only one-fifth the size of Exxon, for example, the company's US refining capacity is more than three-fifths as large as Exxon's. As a refiner, Arco operates over 5% of the nation's capacity, 785,000 bbl/d in 1974 (1). This is more than 50% larger than the next biggest company. The capacity is spread over five refineries in California, Washington, Illinois, Pennsylvania, and Texas. The average size of these refineries, 157,000 bbl/d, is second only to that of Exxon.

Arco's refinery pollution control record to date is relatively good (15). CEP rates both air and water control performance "good." Arco is the only big eight refiner to receive good marks in both areas. The company's overall control of air pollution is the best in the study, and Arco stands third best in water quality, just behind Exxon and Shell. On this basis, we judge that Arco has more of its pollution control costs behind it than the average refiner, and certainly more than companies such as Gulf and Texaco, which are the two worst controlled of the big eight refiners.

CEP estimates that Arco will have to spend $165 million* to buy pollution control equipment between 1974 and 1983; $87 million of this for equipment to complete control at existing refineries, and $78 million to make new capacity environmentally acceptable. The $165 million imposes an extra financing burden of 2-3% over the years to 1983, second only to the impact on Shell's financing needs.† The $87 million for existing capacity comes to $111 for each barrel of existing capacity, the lowest post-1973 per barrel capital requirement among the big eight.

Achieving pollution control above 1973 quality levels will add $270-340 million to Arco's 1974-83 refining costs. More costly, less polluting fuels account for $45 million of the total, and costs related to capital installations are responsible for the rest. These CEP cost estimates translate to an average additional 9-11¢ per barrel of product through 1983, lower than the future costs for any other company except Shell.

In the long run, full control of existing refineries will raise costs 9.9-13.3¢ per product barrel above 1973 cost levels. This study generally focuses on the years through 1983, and that is why we present cost and profit results through that date. CEP cost estimates assume that capital will be installed over a period of years with the final pieces coming on stream in 1983. As a result, average costs for the period 1974-83 will not reflect the full impact of complete control since 1. the average level of control for the period is less than complete, and 2. the assumed life of the equipment is 15 years. The long run figure cited is the cost of complete control over a full equipment life cycle.

Several factors which may modify the cost conclusions do not enter these calculations. These include expansion plans, relative construction costs, present level of low sulfur fuel use, and the largely intangible aspect of the experience and knowledge that a company's present pollution control quality represents.

Refinery expansion is not important to Arco's relative cost position. Arco has as much potential to reduce average company pollution control costs through expansion as the rest of the industry and no more. The company presently has only one major expansion under way, a 50% expansion of the 230,000 bbl/d Houston, Texas refinery (2). Arco had planned to build an entirely new refinery in cooperation with Southern California Edison, but the project has been aban-

*All figures in these profiles are in 1974 dollars.

†We compare the pollution control capital estimate to CEP's estimate for total company investment over the ten years, 1974-83. CEP assumes that a company's 1973 investment level will increase at a compound rate of 4% per year (real terms) through 1983. The resulting figures for each year are then summed to get the total for the period. We use the same methods with the same 4% growth rate to estimate investment totals for each of the eight firms.

doned because of cost factors and supply and demand uncertainties. CEP's assumed expansion rate for Arco roughly mirrors that for the industry up to 1977, and then, using our model, simply matches industry growth through 1983. On this basis we estimate that 26% of Arco's 1983 capacity will have come on stream since 1974, equal to the total industry percentage. Therefore, this is not a source of relatively lower costs for Arco.

Construction cost considerations may raise Arco's relative costs. We calculate that the company's geographic distribution means that its average construction costs will be 5% higher than the industry average, assuming any expansion does not change the geographic pattern (16). Pollution control capital is unlikely to escape the extra costs.

The present level of low sulfur fuel use is an offsetting factor. Arco's record, which is second best in the study, is very little worse than the best, Exxon's. Several Arco refineries seem to be using low sulfur fuels already, and to the extent this is true, company product costs reflect this high cost method of operation. Thus, our estimate of the extra costs of low sulfur fuels may be too high for Arco, compared to the present costs of producing products.

Finally, not only does good current performance mean the company has less of a future job to do, it also means that the firm may be able to complete the control task in a less costly manner than other companies. Pollution control tends to be more an art than a science, a custom designed job from refinery to refinery. Unfortunately much pollution control technique is learned through trial and error, and experience has a very high value in this area. High quality control may mean that a company has valuable experience to draw on and will be able to do a better, cheaper job of completing control than other refiners.

US refining is an important part of Arco's business, and refining cost increases are a serious matter for the company's overall profit picture. CEP calculates that unrecovered costs could reduce 1974-1983 profits by 4-5%, the second most serious impact in this study after Shell. If, however, the industry as a whole achieves an average 75% cost recovery through higher product prices, Arco will be in much better shape. Because its incremental costs are below those of the industry, a 75% average recovery would mean approximately 90% recovery for Arco. The average profit reduction would fall to around .3%, effectively nothing. If the industry were to achieve an average 100% cost recovery, Arco's profit margin would actually rise from 1973 levels.

In a sense, we have described the worst effect that refining pollution control costs can have on Arco. The company has very little capacity outside the United States, and the potential for pollution control costs to make inroads on profits from foreign refineries is trivial. In this context Arco's position is a good one. Unknown foreign variables are not going to enter this picture and confuse the conclusions. This is much less true of the major international companies: Exxon, Gulf, Texaco, Socal, and Mobil.

Company Data Sheet: ARCO

BACKGROUND INFORMATION

General Company and Refinery Information:

'73 Assets	'74 Sales	'74 Profits	'73 Employees	'73 Return on Equity
\$5,109	\$7,167	\$475	26,300	8.9%
Big Eight Rank 1973				
8th	8th	8th	8th	8th

Total U.S. Capacity (bbl/d)	Number of U.S. Refineries	Average Capacity bbl/d	Pollution Control Performance — Water	Pollution Control Performance — Air
785,000	5	157,000	Good	Good
Big Eight Rank 1973				
8th		2nd Largest	3rd Best	Best

CEP PROJECTED RESULTS FOR 1974-1983

POLLUTION CONTROL COSTS
(All figures are in millions of 1974 dollars.)

Capital Investment:

Total	For Existing Capacity	For Existing Capacity cost/bbl	Capital Total as % of 1974-1983 Investment Total
\$169,000,000	\$87,000,000	\$111	2.5%
Big Eight Rank			
Smallest	8th Lowest	8th Lowest	2nd Highest

Product Cost Increase:

Capital Related	Low Sulfur Fuel	Total	Unit Product Cost Increase (¢/bbl) '74-'83 ave.	Unit Product Cost Increase (¢/bbl) Long Run
\$227 (low) \$300 (high)	\$45	\$272 (low) \$345 (high)	9.0 (low) 11.4 (high)	9.9 (low) 13.3 (high)
Big Eight Rank				
			8th Lowest	8th Lowest

CUMULATIVE PROFIT IMPACT

Maximum Reduction: Zero Cost Recovery	"Probable" Reduction: 75% Industry Average Cost Recovery	100% Industry Average Cost Recovery
3.5% (low) 4.5% (high)	.2% (low) .3% (high)	Small Gain
Big Eight Rank		
2nd Highest	2nd Lowest	

POST 1973 POLLUTION CONTROL COST
PER PRODUCT BARREL

[Bar chart showing Exxon vs Industry Average, with 1974-83 Average and Long Run bars, scale 5¢, 10¢, 15¢]

EXXON CORPORATION

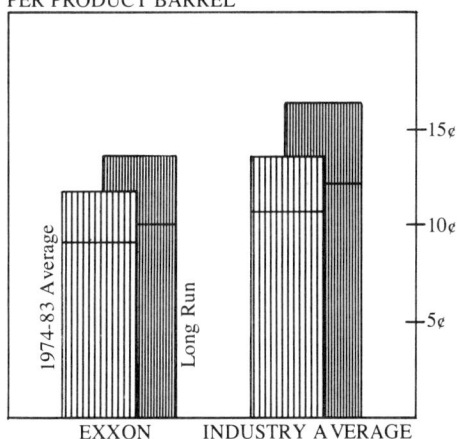

Exxon Corporation is a giant among giants. It is the largest industrial corporation, not only in the United States, but in the world. Exxon's 1974 profits exceeded $3,000 million (17), and sales surpassed $45,000 million, more than the gross national product of all but the largest and most prosperous countries. Texaco, the second largest company CEP studied, is itself the third largest US industrial in terms of assets, but Exxon is nearly twice as large. Most of Exxon's vast operations, however, are outside the United States. Exxon is the classic multinational corporation. While the company is the largest US refiner, it is not as dominant as its international position might suggest. Exxon is so huge that its investment in US refineries amounts to only 4% of the firm's total assets (6).

Exxon's refining capacity in 1974 amounted to 1,252,000 bbl/d, or almost 9% of US refining capacity (1). This is only 10% larger than the next biggest refiner, Shell. All of this capacity is concentrated in only five refineries which average 250,000 bbl/d, by far the largest average size for the big eight. These refineries include the two largest in the country, the 450,000 bbl/d Baton Rouge, Louisiana plant and the 400,000 bbl/d facility at Baytown, Texas.

Exxon's overall pollution control record is among the best of the companies CEP studied (5). Exxon's water pollution control is second only to Shell's and, on the whole, earns a "good" rating. The company still has water pollution problems, but its refineries seem to be making progress toward control and to have gone a good deal further than some of their peers. Consequently, less remains to be done. Exxon's air pollution control record is mixed. Here the company's performance is only "fair," with hydrocarbons and particulates causing special problems. Exxon's SOx control, on the other hand, is the best of the eight refiners. Overall, the firm's air record is the study's third worst.

CEP estimates that Exxon will have to add $145 million worth of new pollution control equipment to complete the control of existing refineries. New capacity will require an additional $133 million. Much of this latter amount will go into expansions at existing refineries, especially the huge Baytown refinery which is adding another 200,000 bbl/d (2). The total pollution control capital spending amount of $278 million is exceeded only by the $322 million Socal will need. Costs per capacity barrel, however, come to only $116, the third lowest amount and well under the estimated industry average of $138 per barrel. The additional financing burden for Exxon over the next ten years is tiny. CEP estimates that refinery pollution control capital requirements will add less than 1% to Exxon's 1974-83 company investment totals.

These capital investment estimates, especially the $145 million for existing capacity, may seem low in light of the $110 million pollution control bond Exxon issued in 1974 (18). The tax exempt bond, issued in April 1974, was specifically earmarked for pollution control at the Baton Rouge refining and chemical operations. The Baton Rouge refinery comprised only about one-third of the company's total domestic capacity at that time. The bond issue, however, cannot be taken as a measure of the net size of the job to be done at that refinery, even after we allow for the fact that some of the $110 million is going for chemical operations. It is likely that much of the bond proceeds will go for equipment that is only partially for pollution control. As Chapter 3 indicates, capital expenditures for pollution control are almost invariably overstated in net terms, and in the case of a pollution control bond, the situation is likely to be even more extreme.

CEP estimates that cleaner refining will add $400-500 million to Exxon's product costs over the years 1974-1983. This includes production from both 1974 existing capacity and assumed expansion. Seventy-five million dollars of this total comes from the use of more costly, less polluting fuels. The remainder is related to the installation of pollution control equipment. These total added costs translate to 9-12¢ per barrel of product, average for the period. Over the long run, pollution control that meets 1983 standards will add 10.2¢-13.8¢ per barrel of product to Exxon's 1973 cost levels, well below the industry average.

Several factors reinforce CEP's conclusion that Exxon's post-1973 pollution control costs should be below industry averages. Some of these factors even suggest that the company's total, from-scratch control costs could be below the $250/barrel average assumed for the industry. First, Exxon's relatively good performance may mean that the company has acquired valuable pollution control experience which could, in turn, mean lower future costs. For example, the Benicia, California refinery profile in *Cracking Down* describes the expensive control mistake Exxon made in deciding to use a single exhaust stack (19). This is at least one error that the company should not repeat.

Second, Exxon is a leader in SOx control. The company reported to CEP that "additional costs have been incurred at some locations due to the use of less polluting fuels (20)." Like Arco, Exxon is already bearing some of the costs of low sulfur fuels, and CEP's estimate for extra costs of these fuels will be too high. Exxon product costs already include some of this element.

Third, Exxon is the only big eight refiner other than Socal with any major expansion actually under way in 1975. In addition to the 50% expansion at Baytown, smaller expansions are in progress at some of the other plants (2). Nonetheless, expanded capacity should account for less than 30% of the company's 1983 total. This is the second largest percentage for the big eight, after Socal, but it is still so small that the possibilities for cheaper control at new capacity are not important to the company's overall cost picture. If de novo control costs 20% less than retrofitting, our estimate of Exxon's average company per barrel costs are only about 3% too high for 1974-83.

Fourth, Exxon's refineries tend to be located in the parts of the country where construction costs are lowest. CEP calculates that average construction costs at Exxon refineries, weighted by size of refinery, come to 95% of the industry average (16). This would imply that, other things being equal, it should be 5% cheaper for Exxon to install pollution control equipment at its existing refineries than it is for the average US refiner.

Post-1973 costs for US refinery pollution control should have little impact on Exxon's total corporate profits. Even if the company had to absorb the full costs, CEP projects that 1974-83 profits would be reduced by only 1%. This is the lowest projected impact for any of the eight companies. The minimal impact results from a combination of Exxon's relatively low added costs and vast size in areas other than US refining. An average 75% industry-wide cost recovery would reduce Exxon's profit damage to nearly zero, and a 100% average recovery would actually result in a small profit increase. Exxon's refining profits have been depressed through 1973 relative to some other firms, in a sense, because the company has already absorbed higher pollution control costs. Alternatively, we can say that other companies have inflated their profits by postponing required expenditures. CEP believes the situation will be reversed by 1983.

Pos-1973 pollution control at US refineries seems to have less significance for Exxon's corporate profits than for any other company in the study. Only one fifth of Exxon's refining capacity is in the United States (6), and pollution control requirements in other countries could be an important problem for Exxon. Overall corporate profits are thus in much more jeopardy from non-US pollution control than for firms such as Arco, Shell and Amoco which have almost all of their refining capacity in the US. On this basis, we judge that Exxon is in no better a position than Arco or Shell, even though the indicated profit impact looks smaller.

Company Data Sheet: EXXON

BACKGROUND INFORMATION

General Company and Refinery Information:

'73 Assets	'74 Sales	'74 Profits	'73 Employees	'73 Return on Equity
	(millions of dollars)			
$25,079	$45,840	$3,140	137,000	18%

Big Eight Rank 1973				
1st	1st	1st	1st	1st

Total U.S. Capacity (bbl/d)	Number of U.S. Refineries	Average Capacity bbl/d	Pollution Control Performance	
			Water	Air
1,252,000	5	250,000	Good	Fair

Big Eight Rank 1973				
1st		1st	2nd Best	6th Best

CEP PROJECTED RESULTS FOR 1974-1983

POLLUTION CONTROL COSTS
(All figures are in millions of 1974 dollars.)

Capital Investment:

Total	For Existing Capacity	For Existing Capacity cost/bbl	Capital Total as % of 1974-1983 Investment Total
$278,000,000	$145,000,000	$116	Under 1%

Big Eight Rank			
2nd Highest		3rd Lowest	Lowest

Product Cost Increase:

Capital Related	Low Sulfur Fuel	Total	Unit Product Cost Increase (¢/bbl)	
(millions of dollars)			'74-'83 ave.	Long Run
$387 (low) $511 (high)	$75	$462 (low) $586 (high)	9.3 (low) 11.7 (high)	10.2 (low) 13.8 (high)

Big Eight Rank				
			3rd Lowest	3rd Lowest

CUMULATIVE PROFIT IMPACT

Maximum Reduction: Zero Cost Recovery	"Probable" Reduction: 75% Industry Average Cost Recovery	100% Industry Average Cost Recovery
.7% (low) .9% (high)	Under .1%	Small Gain

Big Eight Rank		
Lowest	Lowest	

POST 1973 POLLUTION CONTROL COST
PER PRODUCT BARREL

GULF OIL CORPORATION

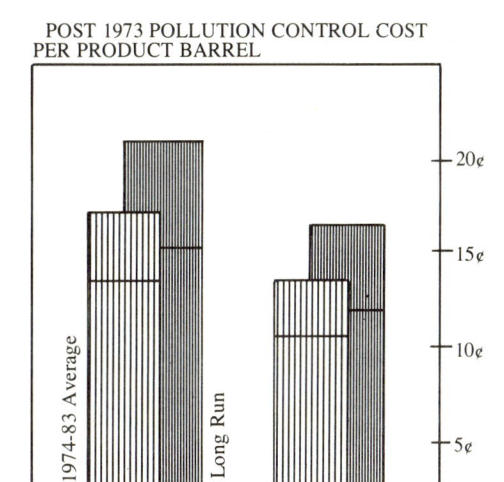

[Bar chart showing Gulf vs Industry Average, with 1974-83 Average and Long Run bars; y-axis scaled from 5¢ to 20¢]

GULF   INDUSTRY AVERAGE

Gulf Oil Corporation, a major multinational corporation, is the study's fourth largest company in terms of assets and fifth largest in terms of sales (13). Gulf's national rank is a good deal more impressive. Its 1973 sales and asset holdings placed the company 11th and 9th respectively on *Fortune's* list of the 500 largest industrials. Among US oil companies, only Exxon and Texaco's profits exceeded Gulf's 1974 net of $1,060 million (21). Approximately 50,000 people worked to earn 14.4% on Gulf's stockholders' equity in 1973, giving it the fifth highest return among the big oil companies and making it number 157 on the list of 500 industrials.

Gulf's position as a refiner is less substantial than its standing as a total company. The company's 860,000 bbl/d capacity, 6.0% of the US total, makes Gulf the country's seventh largest refiner (1). CEP's figures for refinery expansions indicate that Gulf will fall to eighth place by 1977. The company operated eight refineries in 1974 with an average size of 108,000 bbl/d. Among the big eight refiners, only Texaco's refineries are of smaller average size. Actual refinery capacities range from a high of over 300,000 bbl/d at Port Arthur, Texas to a low of 27,000 bbl/d at Hercules, California. The company's large size compared to its US refining capacity status is explained primarily by the scope of its overseas operations. Gulf's domestic refineries constitute only 7% of the firm's total corporate investment (7).

Gulf's overall pollution control record is the worst of the eight companies CEP studied (15). CEP concludes that Gulf has the largest pollution control job ahead of it. Its air quality control earned a "fair/good" fourth best, but its water pollution control, rated "very poor," is the worst of the eight. The Port Arthur refinery, in particular, is one of the study's very worst water pollution controllers. Gulf has a long way to go in the costly area of water pollution control and can draw on little valuable control experience for future control efforts. If the average refiner has installed 60% of the pollution control capital required to meet 1977 goals, we estimate that Gulf has expended only 40-45% of the needed monies.

CEP estimates that Gulf will have to invest $210 million for pollution control over the next ten years; $153 million of this will go to bring existing refineries up to standard. The total amount is only the fifth highest in the study, primarily because Gulf has very little expansion underway or firmly planned. The amount for existing capacity is the largest of any of the companies, although all but Arco have larger capacities. We estimate that Gulf will have to spend $178 per barrel of 1974 capacity, much the largest amount for the eight companies and well above the $138 per barrel average estimate for the industry.

Gulf's capital investment burden comes to approximately 2% of projected total company capital spending for 1974-83 (7), the third largest burden after Arco and Shell. These two latter firms, however, are much more concentrated in the US than is Gulf. The additional amount imposed on Gulf's domestic capital spending is probably the largest in the study.

CEP estimates that Gulf's 1974-83 refining costs will rise by $410-520 million. Of this, $45 million is for more costly fuel, and the remainder is for extra costs related to new capital investments. On a unit product basis, this comes to a 14-17¢ per barrel average for 1974-83, the largest post-1973 cost increase that CEP calculated. In the long run, CEP estimates that Gulf's company refining costs will rise 15.2-20.7¢ per barrel of product above 1973 levels. The graph at the head of this profile shows that these figures are well above the industry average.

Few clear offsets to our cost estimates exist for Gulf, although the company certainly could use tax exempt bond finance as Exxon has done. Gulf's SOx control is reasonably good, which may indicate that the company has already absorbed some of the higher fuel charges. If so, this would reduce incremental costs. The only other factor is a favorable geographical distribution. CEP estimates that Gulf's refineries are so located that company construction costs, including those of pollution control equipment, should come to only 95% of the industry average (16). There seems little other reason to feel our estimates may be too high relative to other firms. Cheaper control from expansion is not a possibility because the company has planned little or no definite expansion. CEP projects that in 1983 only 19% of Gulf's capacity will have been built since 1974. The industry figure will be 26%.

We project a 2-3% decline in Gulf's 1974-83 profits in the unlikely event that no pollution control costs are recovered from customers. This is the fourth lowest decline in the study, primarily because US refining is such a small portion of Gulf's operations. The more probable situation, in which much of the cost increase is passed along to the customer, is somewhat different. We calculate that if the industry recovers an average 75% of 1974-83 pollution control costs, Gulf's profits will be reduced by an average 1%, the heaviest impact of the eight.

Our prognosis for Gulf, then, is the most adverse in the study. We must point out, however, that all this is in comparison to 1973 levels of profit and return on investment. We assume that Gulf's 1973 levels have been inflated relative to other refiners because pollution control costs have been postponed. Full compliance with pollution control standards will reverse this situation. Furthermore, one of the most interesting features of the prognosis for Gulf is that, although this is the worst hit company, the amount involved is really trivial. Even for Gulf, pollution control is not impossibly expensive.

Company Data Sheet: GULF

BACKGROUND INFORMATION

General Company and Refinery Information:

'73 Assets	'74 Sales	'74 Profits	'73 Employees	'73 Return on Equity
	(millions of dollars)			
$10,074	$18,200	$1,060	51,600	14.4%

Big Eight Rank 1973				
4th	5th	3rd	4th	5th

Total U.S. Capacity (bbl/d)	Number of U.S. Refineries	Average Capacity bbl/d	Pollution Control Performance	
			Water	Air
860,000	8	108,000	Very Poor	Fair/Good

Big Eight Rank 1973			
7th		7th	Worst	4th Best

CEP PROJECTED RESULTS FOR 1974-1983

POLLUTION CONTROL COSTS
(All figures are in millions of 1974 dollars.)

Capital Investment:

Total	For Existing Capacity	For Existing Capacity cost/bbl	Capital Total as % of 1974-1983 Investment Total
$210,000,000	$153,000,000	$178	2.1%

Big Eight Rank			
4th Lowest		Most	3rd Highest

Product Cost Increase:

Capital Related	Low Sulfur Fuel	Total	Unit Product Cost Increase (¢/bbl)	
(millions of dollars)			'74-'83 ave.	Long Run
$364 (low)	$45	$409 (low)	13.5 (low)	15.2 (low)
$476 (high)		$521 (high)	17.2 (high)	20.7 (high)

Big Eight Rank			
		Most	Most

CUMULATIVE PROFIT IMPACT

Maximum Reduction: Zero Cost Recovery	"Probable" Reduction: 75% Industry Average Cost Recovery	100% Industry Average Cost Recovery
2.1% (low)	.8% (low)	Small Reduction
2.7% (high)	1.1% (high)	

Big Eight Rank		
4th Lowest	Highest	

MOBIL OIL CORPORATION

POST 1973 POLLUTION CONTROL COST PER PRODUCT BARREL

[Bar chart showing Mobil (1974-83 Average and Long Run) versus Industry Average, with values ranging up to 20¢ per barrel]

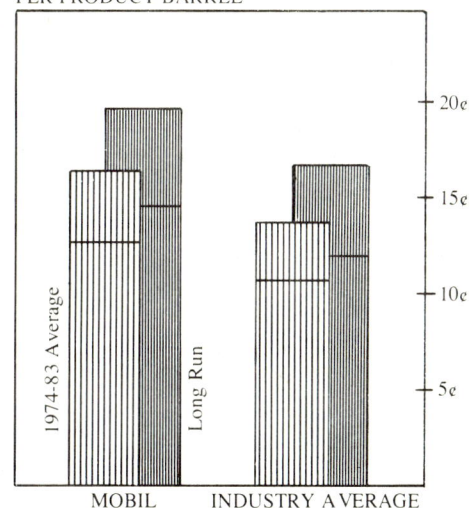

Mobil Oil Corporation, a major multinational company, stands high on the list of largest companies in the United States. Ranked by sales, assets and profits, Mobil was the 6th or 7th largest industrial US corporation in 1973 (13). The firm is also a huge employer. Nearly 74,000 people worked for Mobil in 1973. Profits in 1974 came to $1,040 million (22), fourth largest of the big eight firms and just barely behind third place Gulf. Mobil's $20,370 million sales exceeded those of all the oil companies except Exxon and Texaco, and the company's return on equity in 1973 was nearly 15%, fourth best among the refiners and in 144th place on the *Fortune* 500 list. Results in 1974 should be somewhat better, but not markedly so.

Mobil is the sixth largest US refiner (1). The company operates seven domestic refineries which average 127,000 bbl/d, the fifth largest average size among the big eight refiners. Total company capacity comes to 887,000 bbl/d, 6.2% of national capacity, an amount scarcely larger than seventh place Gulf's. Mobil's US refining is a relatively small part of the company's business. Under 10% of total corporate assets have been committed to US refining (8), and only about one-third of the company's capacity is located in the United States.

Mobil's overall pollution control record is one of the worst of the companies studied (15). CEP rates Mobil's seventh place water control performance "poor." Only Gulf has a worse record. Air control quality earns a "good" description. Only Arco and Socal do better. The company's good air rating, however, is tied to a no emissions performance on CO. SOx control, one of the

79

more costly on a net basis, falls toward the not-so-good end of the scale. On the whole, we conclude that Mobil has more than the average amount of pollution control capital spending left to do. This is especially true in the costly water control area.

CEP estimates that Mobil will need to spend $208 million for pollution control capital from 1974 to 1983. The vast majority of this, $148 million, will go to complete control at existing capacity. Projected company expansion is minimal. This $148 million means that CEP estimates that Gulf will have to spend $167 for each barrel of capacity now in place, an amount exceeded only by Gulf's $178 per barrel. The potential burden on the company's financing activities, however, is still very small. CEP calculates that the $208 million total should add less than 1.5% to Mobil's total capital spending through 1983 (8), the second smallest burden in this study.

Increased pollution control will raise Mobil's costs of refining by $353-462 million through 1983, according to CEP estimates. More costly low sulfur fuels account for $47 million of this, and the rest arises from pollution control capital investment. Extra costs average 13-16¢ per barrel of product over the years 1974-83. The long run cost increase above 1973 levels comes to 14.4-19.6¢ per product barrel. CEP estimates that only Gulf faces higher post-1973 added costs. The graph at the head of the profile indicates how much these costs exceed the industry averages.

Possible cost reducing factors generally seem to go against Mobil. The company has less than the average expansion under way or planned (2). Mobil thus stands to gain less than its competitors from any possibilities of cheaper de novo control. In addition, Mobil's refineries tend to be located in areas where construction costs are relatively high. CEP calculates that the company's construction cost index is 1.05 times the industry average (16). Again the indication is for higher costs.

CEP estimates that the impact on Mobil's profits through 1983 would be one of the least severe in the study in the unlikely event that there is no recovery of costs from customers. This is due primarily to the relatively small part that US refining plays in the firm's total operations. We calculate that the maximum profit reduction is on the order of 2–2.5% over the 1974-83 period. Cost recovery through industry-wide price increases brings this profit reduction down even further, to around 1% total if the industry achieves an average 75% recovery. In this case, however, Mobil's relative position slips. Because projected costs are above industry averages, Mobil stands to regain a smaller percentage of its costs than the industry average from any general price increase. An average 100% cost recovery for the industry still leaves Mobile with very slightly reduced profits. All this, of course, is compared to 1973 profit levels, which we assume have been inflated by the company's postponement of needed pollution control. CEP calculates that Mobil's post-1973 profit reduction, though tiny at 1% (the 75% recovery case), is still larger than all the other companies studied except for Amoco and Gulf.

Company Data Sheet: MOBIL

BACKGROUND INFORMATION

General Company and Refinery Information:

'73 Assets	'74 Sales	'74 Profits	'73 Employees	'73 Return on Equity
	(millions of dollars)			
$10,690	$20,370	$1,040	73,900	14.9%
Big Eight Rank 1973					
4th	3rd	4th	3rd	4th

Total U.S. Capacity (bbl/d)	Number of U.S. Refineries	Average Capacity (bbl/d)	Pollution Control Performance	
			Water	Air
887,000	7	127,000	Poor	Good
Big Eight Rank 1973				
6th		5th	2nd Worst	3rd Best

CEP PROJECTED RESULTS FOR 1974-1983

POLLUTION CONTROL COSTS
(All figures are in millions of 1974 dollars.)

Capital Investment:

Total	For Existing Capacity	For Existing Capacity (cost/bbl)	Capital Total as % of 1974-1983 Investment Total
$208,000,000	$148,000,000	$167	1.4%
Big Eight Rank			
		2nd Highest	7th Highest

Product Cost Increase:

Capital Related	Low Sulfur Fuel	Total	Unit Product Cost Increase (¢/bbl)	
(millions of dollars)			'74-'83 ave.	Long Run
$353 (low)	$47	$400 (low)	12.8 (low)	14.4 (low)
$462 (high)		$509 (high)	16.3 (high)	19.6 (high)
Big Eight Rank				
			2nd Highest	2nd Highest

CUMULATIVE PROFIT IMPACT		
Maximum Reduction: Zero Cost Recovery	"Probable" Reduction: 75% Industry Average Cost Recovery	100% Industry Average Cost Recovery
2.0% (low)	.7% (low)	Small Reduction
2.4% (high)	.9% (high)	
Big Eight Rank		
6th Highest	3rd Highest	

POST 1973 POLLUTION CONTROL COST PER PRODUCT BARREL

SHELL OIL COMPANY

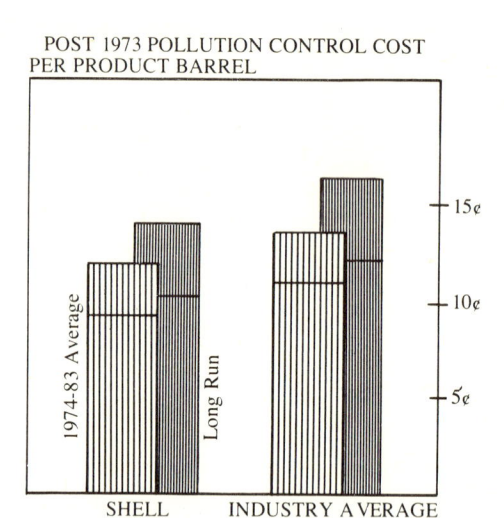

(Bar chart: Shell — 1974-83 Average and Long Run bars; Industry Average bars; y-axis marked 5¢, 10¢, 15¢)

Shell Oil Company is not a large company by integrated petroleum company standards. Its sales, assets and profits barely exceed those of Arco, the smallest firm of the big eight refiners. Shell is small, however, only in the context of the oil industry. The company is an industrial giant, 16th on *Fortune's* 1973 list of the 500 largest industrials when it is ranked by assets and profits (13). Company profits in 1974 came to $621 million (23), more than the *sales,* not to mention the profits, of all but the very largest US corporations.

To say that Shell is small for an oil company can also be misleading. Shell is the partially owned US subsidiary of the giant Royal Dutch Shell group, one of the world's largest corporations (24). Exxon is the only company in this study that is comparable in size to Shell's international parent. We consider Shell here as a separate organization because it operates as one in the United States. The parent company is a majority stockholder, but still about 30% of the subsidiary company's stock is publicly traded, primarily in the United States.

After Exxon, Shell is the second largest US refiner (1). The company operates eight refineries with a total capacity of over 1,125,000 bbl/d, nearly 8% of 1974 US refining capacity. Scattered around the country from California to Delaware, the refineries average 141,000 bbl/d, the third largest average capacity among the big eight refiners.

CEP found that Shell's pollution control record is among the best of the companies studied (15). The firm's water control performance, rated "good," is the best in the study. Its air pollution control earned only a "fair" description.

Shell stands fifth from the top in terms of air quality control. The company seems to have CO under control completely, but has had problems in controlling SOx. Relying primarily on the basis of the company's good water pollution control record, we judge that Shell's existing level of pollution control investment is topped only by Arco's.

CEP estimates that Shell's post-1973 pollution control capital needs will come to $202 million—the second smallest amount for a company in this study—yet the additional pollution control capital will impose a financial burden on Shell that is the highest of the big eight. Existing refinery capacity will take up $128 million of the $202 million total. Shell itself estimates that it spent nearly $40 million for air and water pollution investment from 1970-1973 (25). If we consider the inflation factor and the fact of substantial spending prior to 1970, CEP's estimate appears to be a reasonable order of magnitude compared to Shell's own estimate. The $126 million for existing capacity comes to $112 per barrel, only slightly more than our estimate of $111 per barrel for Arco, the smallest amount in the study. Shell's relatively large financing burden is primarily a function of the proportion of total investment the company has in domestic refining. We estimate that 1974-83 pollution control investment will add around 3% to Shell's total financing needs for that period (9).

Post-1973 pollution control should raise total 1974-83 refinery product costs by $360-460 million according to our estimates. More expensive, cleaner fuel accounts for $60 million, and the rest comes from expenses associated with pollution control equipment. Shell reported to CEP that pollution control operating and maintenance costs were $23 million in 1973 and had been increasing at roughly 10% per year from 1970 through 1973 (25). If this figure is increased to reflect higher control levels and greater capacity and then summed over ten years, it does not look terribly different from CEP's estimate. The total cost figures indicate that Shell faces an average extra cost of 9-11¢ per product barrel through 1983. Long run extras come to 10.1-13.6¢ per barrel above 1973 cost levels. Only Arco does better.

Shell's projected expansion is nominal. The company has been unable to get siting approval for a proposed East Coast refinery. If the company gets approval and goes through with the project, building an entirely new refinery, expansion might increase substantially. This is highly problematical, of course, but it would mean that the company's control costs might be even lower because of cheaper de novo control at the expansion. In addition, if past experience has any value, we can project that Shell's future control costs will be lower than those of other companies because Shell has gotten good on-the-job training. The company has tended to be a leader in control technology, and this may well prove valuable in cost terms.

Pollution control at US refineries hits Shell's profits relatively hard if there is no cost recovery. The company operates almost entirely in this country, and US refineries account for nearly 20% of its corporate assets (9). These pollution control costs are thus more crucial to Shell's profits than they are for most of the other firms studied. CEP projects that maximum profit reduction, when there is no cost recovery, comes to 5-6% over the years 1974-83. Only Arco's 4-5% reduction comes close to this. The picture improves considerably if costs can be passed along to the customer. Our benchmark 75% industry average cost recovery results in Shell's profit reduction falling to only 0.5%. In this case Shell fares better than all its competitors except Arco and Exxon. Better, that is, than several companies which have smaller stakes in US refining. When the industry averages 100% cost recovery, Shell is one of those companies whose profit margins should actually rise above 1973 levels.

Company Data Sheet: SHELL

BACKGROUND INFORMATION

General Company and Refinery Information:

'73 Assets	'74 Sales	'74 Profits	'73 Employees	'73 Return on Equity
	(millions of dollars)			
$5,381	Not Reported	$620	32,080	10.7%

Big Eight Rank 1973					
7th	7th	7th	7th	7th

Total U.S. Capacity (bbl/d)	Number of U.S. Refineries	Average Capacity bbl/d	Pollution Control Performance	
			Water	Air
1,127,000	8	141,000	Good	Fair

Big Eight Rank 1973				
2nd		3rd	Best	5th Best

CEP PROJECTED RESULTS FOR 1974-1983

POLLUTION CONTROL COSTS
(All figures are in millions of 1974 dollars.)

Capital Investment:

Total	For Existing Capacity	For Existing Capacity cost/bbl	Capital Total as % of 1974-1983 Investment Total
$202,000,000	$126,000,000	$112	2.9%

Big Eight Rank			
2nd Lowest		2nd Lowest	Most

Product Cost Increase:

Capital Related	Low Sulfur Fuel	Total	Unit Product Cost Increase (¢1/bbl)	
	(millions of dollars)		'74-'83 ave.	Long Run
$302 (low)	$60	$362 (low)	9.0 (low)	10.1 (low)
$403 (high)		$463 (high)	11.5 (high)	13.6 (high)

Big Eight Rank				
			2nd Lowest	2nd Lowest

CUMULATIVE PROFIT IMPACT

Maximum Reduction: Zero Cost Recovery	"Probable" Reduction: 75% Industry Average Cost Recovery	100% Industry Average Cost Recovery
4.6% (low)	.5% (low)	Small Gain
5.7% (high)	.6% (high)	

Big Eight Rank		
Most	3rd Lowest	

POST 1973 POLLUTION CONTROL COST PER PRODUCT BARREL

STANDARD OIL OF CALIFORNIA

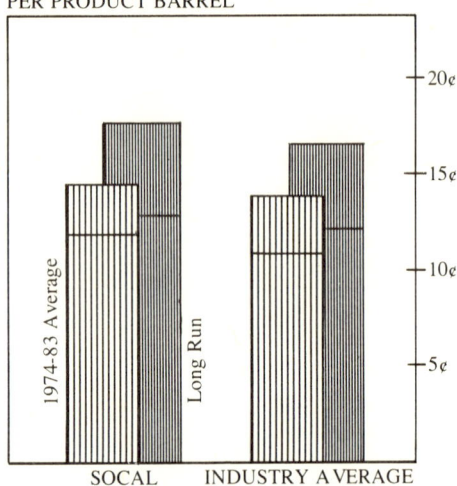

Standard Oil of California (Socal) has far-flung international operations. Most of its crude oil production and over half of the firm's refinery runs were outside the US in 1974 (10). Socal's 1974 sales of $18,800 million and profits of $970 million put it in fourth and sixth place respectively in the big eight rankings (27). In a national context, Socal stood 10th in assets on *Fortune's* 1973 list of the top 500 industrials (13), 11th in sales, and 7th in profits. Return on stockholder's equity, at 14.5%, put the company in 149th place. Results for 1974 will probably move Socal up in all these categories.

The fifth largest US refiner in 1974, Socal's refining ability was approximately 950,000 bbl/d, 6.7% of the US total (1). This capacity was spread among nine refineries, more than any other company except Texaco. The average size of 106,000 bbl/d is the second smallest of all the big eight refiners. Actual sizes range from the tiny 4,000 bbl/d Kenai, Alaska plant, the smallest fuels refinery operated by any of the eight firms, to Pascagoula, Mississippi's 240,000 bbl/d. A major refinery expansion program that is now under way should make Socal the second largest refiner in the country, just behind Exxon, by 1977 (2).

Socal's pollution control performance is mixed (15). The company's air quality results are second best in the study. CEP calls it "good." Only ARCO does a better job. This finding tends to confirm the investment figures the company supplied to CEP (28). Over the last four years, according to Socal, the firm has spent $40 million, or 72% of its total refinery pollution control spending, on air quality. The fact that two of Socal's largest refineries, El Segundo and

Richmond, are located in California encourages this tendency to concentrate on air pollution control. CEP has found that California air quality regulations tend to be relatively strict, and perhaps most important, strictly enforced.

Water quality control at Socal is another story. We find that the company's performance rates only a "fair/poor" description and is third worst in the study. The company's figures for water pollution investment confirm that relatively little effort has gone into this area. Socal reports spending only $15 million on water quality during the years 1970-73. We conclude on the basis of these findings that Socal has a costly pollution control job ahead. Poor performance in the more expensive water control area outweighs the good air quality record, and we find that Socal will have to make more than the average per barrel investment in pollution control.

CEP estimates that Socal will have to invest $322 million in pollution control facilities between 1974 and 1983. This amount is by far the largest total bill for any of the companies in the study. The next largest total amount, Exxon's $268 million, is only 83% of Socal's total. The $322 million would raise Socal's 1974-83 financing requirements by 2% according to CEP calculations (10), the highest burden for any of the multinational firms studied. (Shell's is higher, but Shell is not a multinational firm.) The total amount is high, however, primarily because of Socal's massive expansions. By 1977, the company will expand its capacity by nearly 50% (actually late 1975 if projects come in on schedule), and much of this money is needed to control new capacity. Capital investment to complete control at existing capacity comes to $140 million, the third lowest amount in the study. The $147 cost per barrel of capacity is somewhat above the industry average but well below the highs reached by Mobil and Gulf.

The company itself lists the gross value of pollution control capital in place at $132 million (28); 72% of that is for air control. This figure is somewhat above CEP's implicit estimate of roughly $100 million in place but does not necessarily invalidate our figure. Socal's number undoubtedly includes some elements we would omit because of their returns. This is particularly likely to be true because of the strong influence of air capital in the figure. That portion of pollution control investment has the greatest return.

The products Socal produces after 1973 will cost $510-650 million more to make because of post-1973 pollution control, according to CEP estimates. Low sulfur fuels account for $67 million of this, and the rest arises from increased pollution control investment levels. The individual product barrel costs will rise by a 12-15¢ average over this period and 12.8-17.4¢ for the long run. This ties Socal with Texaco for third highest increased costs, about 7% above the industry average.

Several factors may make these estimates for Socal too high. The most important is the question of expansion. Socal is the only major refiner with substantial expansion in progress (2). Forty per cent of the company's 1983 capacity will have been added since 1974. This means there is a lot of scope for designing good, efficient pollution control into new capacity from the beginning, and that may well bring control costs down. Less polluting fuel costs may be reduced by designs which produce more process gas than can be used for fuel within the refinery, although this would result in greater needs for pollution control capital to desulfurize these gases. CEP has no actual figures on low sulfur fuel use at Socal's refineries, but the firm's standing with respect to SOx control, a strong fourth place showing, indicates the company may be well along the way to already absorbing sulfur reduction costs.

Socal's past refinery profit margins have been inflated by its lagging water pollution control. The company thus faces a relatively large profit reduction, even though the added costs apply to a small portion Jof operations. Socal had only approximately 7% of its corporate assets tied up in US refining as of 1973 (10). This figure will increase as the expansions are completed, but not greatly. If there were no cost recovery, we estimate that cumulative profits over the 1974-83 decade would be reduced 2-3% from the amount they would have achieved if refining costs did not rise from 1973 levels. Only Shell, Arco, Amoco, all much more heavily dependent on US operations, would experience greater reductions. Cost recovery from price increases brings the profit reduction down considerably, but does not improve Socal's relative standing. At an industry average recovery of 75%, Socal's 1974-83 profits will fall by about 1%. Only Gulf and Amoco do worse, and only slightly worse at that. One hundred per cent average industry cost recovery would still leave Socal with profit margins slightly depressed from 1973 levels.

Company Data Sheet: SOCAL

BACKGROUND INFORMATION

General Company and Refinery Information:

'73 Assets	'74 Sales	'74 Profits	'73 Employees	'73 Return on Equity
	(millions of dollars)			
$9,082	$18,800	$970	39,269	14.5%
Big Eight Rank 1973					
5th	4th	6th	6th	2nd

Total U.S. Capacity (bbl/d)	Number of U.S. Refineries	Average Capacity bbl/d	Pollution Control Performance	
			Water	Air
952,000	9	106,000	Fair/Poor	Good
Big Eight Rank 1973				
5th		7th	3rd worst	2nd Best

CEP PROJECTED RESULTS FOR 1974-1983

POLLUTION CONTROL COSTS
(All figures are in millions of 1974 dollars.)

Capital Investment:

Total	For Existing Capacity	For Existing Capacity (cost/bbl)	Capital Total as % of 1974-1983 Investment Total
$322,000,000	$140,000,000	$147	2.0%
Big Eight Rank			
Most		5th Highest	3rd Highest

Product Cost Increase:

Capital Related	Low Sulfur Fuel	Total	Unit Product Cost Increase (¢/bbl)	
(millions of dollars)			'74-'83 ave.	Long Run
$444 (low)	$67	$511 (low)	11.5 (low)	12.8 (low)
$581 (high)		$648 (high)	14.6 (high)	17.4 (high)
Big Eight Rank				
			3rd Highest	3rd Highest

CUMULATIVE PROFIT IMPACT

Maximum Reduction: Zero Cost Recovery	"Probable" Reduction: 75% Industry Average Cost Recovery	100% Industry Average Cost Recovery
2.5% (low)	.7% (low)	Small Reduction
3.2% (high)	.9% (high)	
Big Eight Rank		
4th Highest	4th Highest	

POST 1973 POLLUTION CONTROL COST
PER PRODUCT BARREL

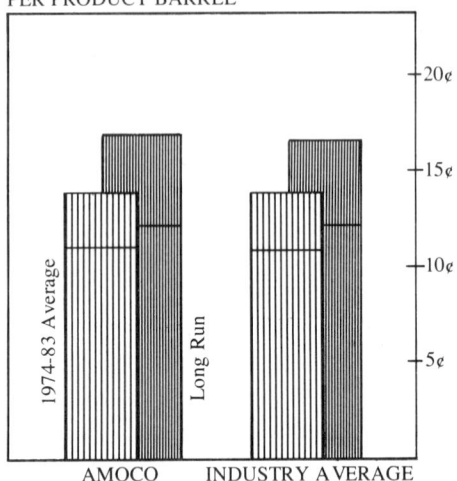

STANDARD OIL COMPANY
(INDIANA)

[Bar chart showing Amoco vs Industry Average, with 1974-83 Average and Long Run values, scale 5¢ to 20¢]

AMOCO   INDUSTRY AVERAGE

Standard Oil Company (Indiana), otherwise known as Amoco, is the largest of the three domestically oriented companies in this study. Its 1974 sales of $10,200 million put it in sixth place among the eight oil companies, and its $970 million 1974 profits tie with Socal for fifth place (26). Arco and Shell, the other two non multinationals, are only about three-quarters as large. Amoco stood 12th in assets among all US industrials in 1973, 15th in sales and 13th in profits (13). The firm's sales and profits positions should move up in 1974. The company's 46,589 employees made it the fifth largest oil company employer and the 81st largest US industrial employer in 1973.

Amoco's 1974 refining capacity, 1,043,000 bbl/d, was 7.3% of 1974 US capacity and third largest in this study (1). A total of eight refineries, with an average size of 130,000 bbl/d, are scattered all over the eastern and central regions of the country. Four of Amoco's refineries are small, under 50,000 bbl/d, and are likely to be relatively costly to control. Over half the company's capacity is concentrated in two 300,000 bbl/d giants, Whiting, Indiana and Texas City, Texas.

Amoco's pollution control performance is toward the bottom end of the rankings (15). Although the company earned "fair to good" marks for its fourth place in water quality, air control is "poor," making Amoco the second worst air performer. The company has real problems with particulate and SOx control, the two areas of air pollution that are most costly to deal with on a net basis. Because Amoco handles the expensive water pollution problem somewhat better than the average firm, however, this offsets the air problem to a degree. We conclude that the company's future pollution control capital needs will approximately equal the industry average.

CEP estimates that Amoco will have to finance $214 million of pollution control capital from 1974 to 1983, the fifth largest total amount in the study. Our projected expansion for Amoco is relatively low, and the bulk of the capital total, $144 million, is to complete compliance at existing refineries. This comes to $138 per barrel, equal to the refining industry average and the fifth largest amount for the big eight firms. The total pollution control capital bill will add nearly 2% to Amoco's financing needs through 1983 (11), the fifth highest burden among the eight firms.

Product costs summed over the years 1974-83 will increase by $400-510 million because of improved pollution control. Some $55 million of this total is to purchase less polluting fuels. The remainder goes to own, operate, and maintain the pollution control capital investment. Costs per product barrel equal the industry rate, an 11-14¢ average through 1983 and 12.1-16.4¢ per barrel in the long run. Four companies face higher post-1973 costs and three confront lower charges.

Amoco announced major expansion plans in the summer of 1974 (2), but nothing appears to have been done since the announcement. The uncertainties and questions that confront all petroleum refiners plague Amoco as well. We conclude that Amoco does not appear to have immediate prospects for expansions of any great size. This means that there is little scope for cost reduction through cheaper de novo control.

Two other factors also argue that our estimate for Amoco's costs is not overstated. The company's poor SOx performance may mean it will require the full additional amount for less polluting fuels. Second, the firm's geographical distribution is such that it faces higher than average construction costs. We calculate that Amoco's average construction cost at existing refineries should run 4% higher than industry average costs for fundamentally similar investment projects (16). In addition, Amoco operates a number of small refineries which may be more costly to control on the average.

Domestic refining is an important part of Amoco's business, accounting for 12% of corporate assets (11). Eighty-five per cent of the company's total refining capacity is in this country. Consequently, cost increases in US refining can have an important impact on total corporate profits. We calculate that unrecovered costs would reduce Amoco's 1974-83 profits by 3-4%, the third most severe impact in the study after Shell and ARCO. Cost recovery from industry-wide price increases would bring this profit impact down. Since Amoco's costs equal the industry average, an X% cost recovery for the industry would make an X% recovery for Amoco.

Our benchmark probable cost recovery of 75% brings Amoco's profit reduction down to about 1%. While this is still the second worst impact in the study, it is really a trivial amount. Full industry cost recovery would mean that pollution control would not damage Amoco's profit margins at all.

Company Data Sheet: AMOCO

BACKGROUND INFORMATION

General Company and Refinery Information:

'73 Assets	'74 Sales	'74 Profits	'73 Employees	'73 Return on Equity
(millions of dollars)				
$7,081	$10,200	$970	46,589	12.9%

Big Eight Rank 1973					
6th	6th	5th	5th	6th

Total U.S. Capacity (bbl/d)	Number of U.S. Refineries	Average Capacity bbl/d	Pollution Control Performance	
			Water	Air
1,043,000	8	130,000	Fair/Good	Poor

Big Eight Rank 1973				
3rd Highest		4th Highest	4th Best	2nd Worst

CEP PROJECTED RESULTS FOR 1974-1983

POLLUTION CONTROL COSTS
(All figures are in millions of 1974 dollars.)

Capital Investment:

Total	For Existing Capacity	For Existing Capacity cost/bbl	Capital Total as % of 1974-1983 Investment Total
$214,000,000	$144,000,000	$138	1.9%

Big Eight Rank			
4th Highest		5th Highest	5th Highest

Product Cost Increase:

Capital Related	Low Sulfur Fuel	Total	Unit Product Cost Increase (¢/bbl)	
(millions of dollars)			'74-'83 ave.	Long Run
$347 (low) $454 (high)	$55	$402 (low) $509 (high)	10.8 (low) 13.7 (high)	12.1 (low) 16.4 (high)

Big Eight Rank			
		5th Highest	5th Highest

CUMULATIVE PROFIT IMPACT

Maximum Reduction: Zero Cost Recovery	"Probable" Reduction: 75% Industry Average Cost Recovery	100% Industry Average Cost Recovery
3.3% (low) 4.1% (high)	.8% (low) 1.07% (high)	0

Big Eight Rank		
3rd Highest	2nd Highest	

POST 1973 POLLUTION CONTROL COST
PER PRODUCT BARREL

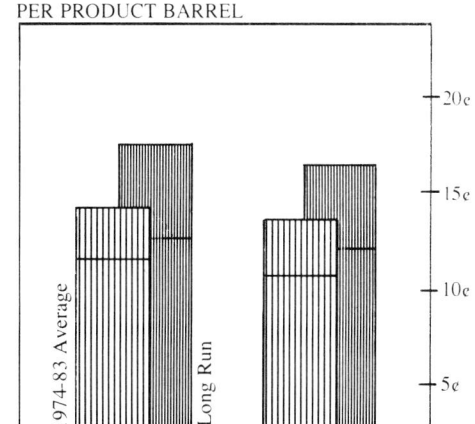

TEXACO

[Bar chart with y-axis marked 5¢, 10¢, 15¢, 20¢; bars labeled "1974-83 Average" and "Long Run" for TEXACO and INDUSTRY AVERAGE]

TEXACO   INDUSTRY AVERAGE

Texaco, a major multinational firm, is the second largest US oil company (13). Only Exxon is bigger. The company sold $24,990 million worth of products in 1974 (29), or approximately $7 worth for every man, woman, and child in the world. Profits came to $1,590 million. Both these figures are second highest among the oil companies. Texaco stood third in terms of assets, fourth in profits and sixth in sales on *Fortune's* 500 of 1973 (13). Long one of the most profitable oil companies, Texaco enjoyed a 16.2% return on equity in 1973, second best for the petroleum companies and 99th best among the industrials. Results in 1974 should be even better.

Texaco is the largest gasoline marketer in the United States (30) and the only oil firm to operate in every one of the 50 states. The company is the fourth largest US refiner, just behind third place Amoco and second place Shell. Total 1974 capacity came to 1,037,000 bbl/d, 7.3% of the US industry total (1). The company operates 11 finished fuel refineries, the largest number of such refineries operated by any single refining company in the US. Their average size, 94,000 bbl/d, is much the lowest of the big eight refiners, and the plants are widely scattered geographically. Even the small average size does not indicate how small Texaco refineries are. The company operates one 400,000 bbl/d giant at Port Arthur, Texas. Nine of the remaining ten have capacities well below 100,000 bbl/d. This production pattern presumably reflects Texaco's 50-state marketing efforts since it is generally cheaper to transport crude oil than finished products.

In some respects Texaco can be rated the study's worst pollution controller (15). The company's air quality performance is "poor," the worst among the big eight. Water quality, fifth best, rates only "fair." The air pollution control problems are puzzling because they occur in areas where the returns to control can be substantial. Texaco's CO performance, for example, is far and away the worst in the study, but this pollutant can be controlled quite cheaply or even at a profit. The same is true of hydrocarbon emissions, and again Texaco is the worst offender. On balance, we conclude that Texaco will have to put more pollution control capital in place than the average refiner, but less than Gulf and Mobil.

CEP estimates that Texaco will have to invest $227 million in pollution control equipment from 1974 to 1983, the third largest amount in the study after Socal and Exxon. Most of the total, $152 million, goes to complete control at existing refineries. This averages $147 per barrel, the same amount that Socal will have to spend, and higher than all other firms except Gulf and Mobil. CEP calculates that $227 million would add only 1.4% to Texaco's total corporate capital spending during this period (12), a relatively small burden compared to other major refiners.

Improving pollution control from 1973 levels will increase Texaco's production costs for 1974-83 products by $500-635 million; $66 million of that is for low sulfur fuels. The costs per product barrel average 12-15¢ for the 1974-83 period and come to 12.8-17.4¢ per barrel over the long run. These figures exceed industry averages by approximately 7%.

Texaco has a minimal amount of expansion in progress. The company had planned a major expansion at the Convent, Louisiana refinery (2), but the project has been shelved for the time being because of uncertainties about government energy policies (31). If it does go through, Texaco will be in a position to enjoy some average cost reduction from de novo control. On a more positive note, Texaco's refineries are located in a geographical pattern which results in relatively low average construction costs. We calculate that Texaco's average refinery costs come to only 95% of the industry average (16). This means that pollution control installations at Texaco's refineries may be somewhat less costly than those at other big eight refineries.

Despite the relatively high post-1973 costs for Texaco, total corporate profit reduction would be only 1.6-2.0% if no costs are recovered. Only Exxon fares better in the no recovery situation. While Texaco is a very big refiner in the US, domestic refining is a small part of the firm's overall business. CEP estimates that Texaco has only 6-7% of corporate assets committed to US refineries (12). This is the major reason that US pollution control does not greatly damage the firm's profits. Texaco does not fare quite so well when there is industry-wide cost recovery. Because the company's costs exceed the industry average, any given industry recovery level will mean a somewhat smaller per cent recovery for Texaco. Even so, with an average 75% cost pass through for the industry, we calculate that Texaco's profit reduction falls to the 0.5-0.6% range, essentially nothing.

Company Data Sheet: TEXACO

BACKGROUND INFORMATION

General Company and Refinery Information:

'73 Assets	'74 Sales	'74 Profits	'73 Employees	'73 Return on Equity
	(millions of dollars)			
$13,595	$23,990	$1,590	74,918	16.2%

Big Eight Rank 1973				
2nd	2nd	2nd	2nd	3rd

Total U.S. Capacity (bbl/d)	Number of U.S. Refineries	Average Capacity bbl/d	Pollution Control Performance	
			Water	Air
1,037,000	11	94,000	Fair	Poor

Big Eight Rank 1973				
4th		8th	4th Best	Worst

CEP ESTIMATE FOR 1974-83

POLLUTION CONTROL COSTS
(All figures are in millions of 1974 dollars.)

Capital Investment:

Total	For Existing Capacity	For Existing Capacity cost/bbl	Capital Total as % of 1974-1983 Investment Total
$227,000,000	$152,000,000	$147	1.4%

Big Eight Rank			
3rd Highest		3rd Highest	6th Highest

Product Cost Increase:

Capital Related	Low Sulfur Fuel	Total	Unit Product Cost Increase (¢/bbl)	
(millions of dollars)			74-83 Ave	Long Run
437 (low)	$66	$503 (low)	11.5 (low)	12.8 (low)
$569 (high)		$635 (high)	14.6 (high)	17.4 (high)

Big Eight Rank				
			6th Highest	3rd Highest

CUMULATIVE PROFIT IMPACT

Maximum Reduction: Zero Cost Recovery	"Probable" Reduction: 75% Industry Average Cost Recovery	100% Industry Average Cost Recovery
1.6% (low)	.5% (low)	Small Reduction
2.0% (high)	.7% (high)	

Big Eight Rank		
7th Highest	5th Highest	

REFERENCES

1. Aillen Cantrell, "Annual Refining Survey," *Oil and Gas Journal*, 1 April 1974, pp. 82-110.
2. "World-Wide HPI Construction Boxscore," *Hydrocarbon Processing*, February 1974; "Worldwide Construction, Refining,"*World Directory, Refining and Gas Processing, 3rd Edition* (Petroleum Publishing Company 1974); and various refining company press releases.
3. P. J. Beall and J. M. Seamans, Texaco, Inc., personal communication, 13 January 1975.
4. Committee on Interior and Insular Affairs, U.S. Senate, "Measurement of Corporate Profits," Serial #93-39 (92-74), March 1974, p. 11.
5. Atlantic Richfield Company, annual reports to stockholders for 1972, 1973; statistical supplements to annual reports for 1972, 1973.
6. Exxon Corporation, annual reports and statistical supplements for 1972, 1973.
7. Gulf Oil Corporation, annual reports for 1972, 1973; Financial and Statistical Supplement for 1972; and Gulf Fact Sheet 1973-1974.
8. Mobil Oil Corporation, annual reports for 1972 and 1973.
9. Shell Oil Company, annual reports for 1972 and 1973.
10. Standard Oil Company of California, annual reports for 1972 and 1973.
11. Standard Oil Company (Indiana), annual reports for 1972 and 1973; and Form 10-K for 1973.
12. Texaco, Inc., annual reports for 1972 and 1973, and Statistical Supplement to 1973 Annual Report.
13. "*Fortune's* Directory of the 500 Largest Industrial Corporations," *Fortune*, May 1974, pp. 230-255.
14. *Wall Street Journal*, "Earnings Digest," 4 February 1975.
15. Gregg Kerlin and Daniel Rabovsky, *Cracking Down: Oil Refining and Pollution Control* (New York: Council on Economic Priorities, 1975).
16. CEP calculation, based on data in Brown and Root, Inc., "Economics of Refinery Wastewater Treatment," API Publication #4199, prepared for the American Petroleum Institute, Committee on Economic Affairs, August 1973.
17. *Wall Street Journal*, "Earnings Digest," 24 January, 1975.
18. Bond Prospectus dated 23 April 1974, Parish of East Baton Rouge, Louisiana: $110,000,000 @ Pollution Control Industrial Revenue Bonds, 5.90%, Series A, Due May 1, 1999. Payment of principal and premium, if any, and interest on bonds is guaranteed by Exxon Corporation.
19. Kerlin and Rabovsky, refinery profile, Exxon, Benecia, California.
20. Exxon Corporation, response to CEP's economic and financial questionnaire.
21. *Wall Street Journal*, "Earnings Digest," 31 January 1975.
22. *Ibid.*, 29 January 1975.
23. *Ibid.*, 28 January 1975.
24. James Ridgeway, *The Last Play* (New York: E. P. Dutton & Co., Inc., 1973), p. 284.
25. Shell, response to CEP economic and financial questionnaire.
26. *Wall Street Journal*, "Earnings Digest," 24 January 1975.
27. *Ibid.*, 29 January 1975.
28. Standard Oil of California, response to CEP economic and financial questionnaire.
29. *Wall Street Journal*, "Earnings Digest," 29 January 1975.
30. Albert J. Fritsch and John W. Egan, *Big Oil: a Citizen's Factbook on the Major Oil Companies*, Center for Science in the Public Interest, 1973.

APPENDIX

Part I of this appendix describes the way CEP developed its estimate of refining industry pollution control capital requirements. Part II presents the model CEP used to convert capital estimates to a product cost basis. Many elaborate details and theoretical niceties have been omitted, but we will be happy to discuss them with the interested reader.

Part I: CEP's Capital Estimate

CEP's estimate is based on information about pollution control's share of total refinery costs. Estimates made in 1973 indicated that approximately $2100 for each barrel of daily capacity would be required to build a large, environmentally acceptable refinery with a wide product range (1). CEP's adjustment to a 1974 dollar basis raises this figure to $2300. Several sources, including some of the companies CEP studied, provided figures for the pollution control portion of this total (2,3,4,5). Most of these figures range from 8-15%, although some sources reported as much as 20-30% add-on cost for pollution control.

CEP believes the very high estimates include a large element of overstatement. The high estimates probably include the full cost of equipment which is only partially or even marginally for pollution control, and they may include capital to produce environmentally acceptable products (6). For the purposes of our capital estimate, CEP assumes that the pollution control needed to meet the requirements of the Clean Air Act of 1970 and the water quality regulations that become effective in 1977 adds an average 10%, or $210 per barrel,* to new

---

*The TOTAL cost of the environmentally acceptable refinery is $2,300 per barrel, including pollution control equipment. We assume that pollution control adds 10% to the cost of an uncontrolled facility. Algebraically, we calculate the control portion as follows: Let X = cost of uncontrolled facility, 10% X the cost of pollution control. Then total cost is X + 10%X = $2,300/barrel. Uncontrolled cost then equals approximately $2,100/barrel and pollution control approximately $210/barrel.

refinery investment. The cost of retrofitting controls to existing refineries would generally be higher, perhaps by as much as 10-20% (7,8).

National refining capacity was approximately 14,200,000 bbl/d at the beginning of 1974 (9). Most of this capacity is at least partially depreciated (10), and many existing refineries do not produce the most costly products. Since the 10%, $210 per barrel, estimate is based on a new refinery with a full product range, applying this figure to existing capacity yields a conservative estimate for capital costs, despite the higher costs of retrofitting. On the 10% basis, it would take a total of $3,000 million to bring existing capacity from no control at all to a level adequate to meet air and water pollution control requirements. However, much pollution control equipment is already in place.

American Petroleum Institute (API) survey figures indicate that the US manufacturing arms of petroleum companies installed $1,700 million worth of pollution control capital between 1966 and 1972 (11). The companies included in the survey represented 88% of national refining capacity in these years, and refining covers approximately 65% (12) of petroleum company investment in manufacturing. The rest is in chemical operations. From these factors, we estimate the API survey figures show that the domestic refining industry installed $1,300-$1,500 million worth of such equipment from 1966 to 1972. Federal government survey figures for 1973 add $300 million to this total (5).

Neither the API survey numbers nor those of the US Commerce Department are corrected to 1974 dollars, nor do they indicate the amount of refinery pollution control capital that was installed before 1966. Uncorrected use of these numbers thus substantially understates monies already spent compared to needs calculated on the basis of 1974 costs. Making some small allowance for these factors, CEP heroically calculates that the remaining capital requirements to upgrade existing capacity are on the order of $1,000 million.

New refinery capacity will also need pollution control equipment. CEP assumes that the industry will grow at an average 3.5% each year over the period 1974-83, and that the pollution control capital needed to meet 1977 standards will cost $210 per barrel of daily capacity. The 1974-77 bill, then, comes to $450 million to control an additional 2,100,000 bbl/d capacity.

CEP assumes that air quality standards will have been met by 1977, and that existing refineries will require no additional air pollution control equipment. The same assumption cannot be made about water pollution control. Water standards for 1983 are considerably more demanding than those of 1977. Since the last units of pollution are more difficult to control, more thorough treatment, which is correspondingly more expensive, is required (13). The cost per unit reduction in pollutants from 1977 to 1983 will be much more than unit costs to reach 1977 standards, but it is difficult to tell how much more expensive. Some of the control methods that will be used are now unknown or in the experimental stages. Estimates based on currently available technology are very likely to overstate costs.

CEP assumes that meeting more severe 1983 water standards will raise refinery pollution control costs to 12% of total construction cost, or around $250 per barrel in 1974 dollars. While we intend this figure only as an order of magnitude and not as an accurate value, it does indicate a substantial increase in water pollution control costs over 1977 levels. At the 12% rate, added capacity of 3,750,000 bbl/d (3.5% growth) will require approximately $1,000 million in pollution control capital. Additional spending to upgrade existing capacity would come to $900 million, meaning that a total of $1,900 million more pollution control capital will be needed from 1978 to 1983.

TABLE A-1

CEP Estimates: Pollution Control Capital Investment Needs for the Petroleum Refining Industry, 1974-1983

(All figures are in millions of 1974 dollars.)

	1974-77	1978-83	TOTAL
To Upgrade Existing 1974 Capacity	1,000	800	1,800
To Control New 1974-77 Capacity	450	100	550
To Control New 1978-83 Capacity		1,000	1,000
TOTAL	1,450	1,900	3,350

Part II: Conversion of Capital Estimates to Product Costs

Capital spending requirements are not directly applicable to product costs because capital equipment is used over a period of years. We have used a simple model to calculate how much product costs rise because of the installation and ownership of pollution control capital. The capital related portion of the pollution control costs has three major components: depreciation, cost of capital or interest charge, and operating and maintenance expenses. Capital related product costs is simply the sum of these three:

Depreciation + Cost of Capital + Operating and Maintenance

We include a sample calculation at the end of Part II to illustrate how our model, described in the following pages, actually functions in practice. Line references in the model description refer to corresponding lines in the sample calculation.

The Base Capital Figure, the Time Frame, and Timing

This study is concerned with how much product costs will change over 1973 levels, not with the full costs of pollution control. The full costs would include some amounts which are already reflected in 1973 costs because there is already a fair degree of control at most refineries. Consequently, the base capital amount for calculations of cost per barrel of product is that amount which will give us post-1973, capital related costs. That amount is simply the new capital required to complete control at existing refineries. CEP assumes the industry's total 1974-83 pollution control capital investment will be $3,500 million. This figure is a bit higher than our estimate and lower than Brown and Root's. Of this, nearly $2,000 million is for control of old refineries. Total dollar costs at all refineries during the period is the product of the per barrel costs and assumed production levels, from both new and old refineries, for the period in question.

CEP has focused on the years 1974-83 when control regulations are coming into force. Hence, we have calculated the added costs through those years, taking the pollution control level of 1973 as given. The calculated figures indicate how much costs will rise, on the average, for the period. In addition, we have made a supplementary calculation of the long run costs per barrel of product. This includes the cost of owning, operating, and maintaining the pollution control equipment over its full life cycle (assumed to be 15 years). The primary difference, then, for the long run estimate is that the full 15 year life of the equipment comes into play. The 1974-83 period includes only two-thirds of the assumed life cycle. This appendix focuses on this latter ten year period.

All aspects of capital related costs are affected by the timing of the capital investment. The sooner the investment is made, the sooner the costs arise. This is relevant when the equipment is phased in over a number of years, as we assume it will be, and when the period in question is less than the assumed life of the equipment. Consequently, the first step in translating the base capital figure to added product costs is to assume a timing pattern. Basically, there are two periods, 1974-77 and 1977-83. For our cost estimate we assume that $1,300 million of the nearly $2,000 million total will fall within the first period, and the remainder will come in the later years. For want of any more soundly based method, the base capital amount is spread evenly within each period. Next, we assume that the full amount of the capital investment made in each year is installed at the beginning of the year. Since the allocation of the total to individual years is somewhat arbitrary, estimates for any specific year have much less meaning than the estimated average cost for the overall period. Such single year estimates are not even presented in the text of the book.

Depreciation (line 2)

CEP assumes that pollution control capital has an average fifteen year life. This figure is something of a compromise. The Internal Revenue Service guideline life for depreciable assets used in petroleum refining is 16 years (14). Brown and Root used 15 years for water pollution control equipment, and the EPA estimates a 10 year life for air pollution control equipment (13,8).

Our calculations use the straight line depreciation method. The individual firms CEP studied use several methods, but straight line dominates, at least for stockholder reporting purposes (15). The total period results are not terribly sensitive to the method chosen. For example, costs calculated on a straight line basis come to 99% of costs calculated on a double declining balance basis. This result obtains because we are not doing a present value analysis. Accelerated depreciation certainly gives less costly results on the present value basis since it results in better cash flow and lower present value of taxes.

The 1969 Tax Reform Act provided an optional accelerated depreciation for some pollution control equipment, an option valuable for cash flow. What the Congress gave with one hand, however, it took away with the other. Possible uses of the option are quite limited, and usage eliminates the investment tax credit for the equipment in question in most cases. Thus, the net value of the accelerated depreciation is questionable (16). At least one petroleum company, however, has used the optional depreciation. Shell informed CEP that it has used the method, but added, "The election of [five year depreciation] did afford us an immediate though minor cash flow advantage; however, the long-term results are inconclusive (4)." CEP has not used this option in its calculations because of its very limited application and the uncertainties about its actual value.

The assumption that capital investments are always made at the beginning of the year means that depreciation for a given year will be calculated on a base that includes equipment installed during that year.

Cost of Capital (lines 3 through 7)

The cost of capital is the cost to a company when funds are tied up in capital investment. This is basically an opportunity cost element, i.e., a measure of the money the funds could have earned in the best available alternative investment project. We calculate the rate of foregone earnings by estimating the return the company would have to earn to raise the necessary funds. CEP assumes this required rate, on a pre-tax basis, comes to 25%. Such a figure does not reflect the costs of a single financing operation, for example a new bond issue. Instead it is based on a consideration of a weighted average cost of capital, as follows:

$$\text{Cost of Capital} = \frac{\text{Debt}}{\text{Debt} + \text{Equity}} (\text{Cost of Debt}) + \frac{\text{Equity}}{\text{Debt} + \text{Equity}} (\text{Cost of Equity})$$

CEP assumes:
> 1) Capital structure is 25% debt, 75% equity, fairly similar to typical structure in the industry. There is no preferred stock or other capital.
> 2) Pre tax cost of debt is 10%. This includes underwriting and other costs as well as overt interest.
> 3) Pre tax cost of equity is 30%. This translates to roughly a 15% after tax return on marginal investments. This percentage is often cited as the return needed to attract new capital to the industry (1).

In numbers the equation thus becomes:

$$\text{Cost of Capital} = (25\%)(10\%) + (75\%)(30\%) = 25\%$$

A lower cost of debt or higher debt/total capital ratio would raise the return on equity implicit in a constant 25% cost of capital figure.

Returns on book equity for most of the petroleum companies in this study were equal to or above 15% in 1973 (15). Results in 1974 also seem to be in this range or higher (17). Historical rates for the industry, however, are considerably lower, and the return for the refining industry alone should be expected to be lower than for the total petroleum industry because refining is generally less risky than the petroleum business as a whole (18). In addition, book equity is probably understated relative to the real value of assets, so opportunity cost returns are below the stated returns. Consequently, CEP considers that the 25% cost of capital for refining yields a conservative estimate of product cost.

This 25% cost of capital is assessed only against net investment, total capital less the 7% investment tax credit (ITC) and cash flow from depreciation. Depreciation is described above. The ITC is a tax benefit generally available to business investment. As of 1974, business taxpayers could reduce their federal income tax bill in a given year by 7% of the cost of certain eligible capital investments made in that year. Changes in the credit are pending in Congress but, as of this writing, have not taken final form. We assume that the credit will remain in force, and since the net change in value of the proposed changes is unclear we make the simple assumption that the 7% rate and present form of the law is still applicable. We then assume that all pollution control capital investment in this industry will be eligible for the credit and that the companies will be able to take full advantage of it. The result is a reduction in the net cost of capital equipment. On a pre tax basis, the 7% credit reduces capital investment costs by 13.46%, assuming the 48% marginal tax rate applies (see "Taxes and Other Issues" below for a discussion of the relevant tax rate).

CEP assumes that depreciation cash flow is evenly distributed over the year, and that the ITC is taken immediately upon installation of the equipment. Cost of capital is then charged against net value of capital in place at the beginning of the year, including new capital investment for the year, less the ITC from that capital and less one half the total depreciation incurred in that year. In equation form, net capital for any given year is:

$$\left[\sum_{1}^{t=n}(C_t - ITC_t)\right] - \left[\sum_{1}^{t=n} D_{t-1}\right] - \tfrac{1}{2}D_n$$

where C_t is capital installed in year t.
ITC_t is the investment tax credit on C_t.
D_t is depreciation charged in year t.

Cost of capital is 25% of the amount given by this equation.

Operating and Maintenance Expense (lines 8 through 10)

Like all capital equipment, pollution control capital requires annual expenditures to operate and maintain. The cost figures which several companies supplied to CEP provide some guidance to the amount of money involved, but the accounting basis is unknown (4,19,20,21). This is especially true of the assignment of joint costs. In the absence of good, specific dollar estimates for these expenditures, CEP assumes that operating and maintenance costs will amount to 10-20% of gross cumulative capital installed. We calculated cost figures for both ends of this range.

The EPA and Brown and Root both use lower operating and maintenance estimates (7,8,13). Figures as high as 25% to 30% of gross capital can be obtained, however, if one assumes that operating and maintenance costs will bear the same relationship to pollution control capital that total refinery costs bear to total refinery investment (22). Such an assumption probably yields too high an estimate for pollution control. Much control equipment is maintained and operated incidentally to ordinary process equipment and requires very little additional effort. Some water pollution control facilities, in particular, have low operating costs. The CEP range is a compromise in an uncertain situation.

Net Increase in Capital Related Product Costs (lines 11 through 16)

Calculated costs per unit of product will vary directly with the product output assumption. The capital related product cost increase is the sum of depreciation, cost of capital, and operating and maintenance expense. As calculated, all these costs are fixed, unrelated to production levels. This is appropriate for depreciation and cost of capital but not entirely accurate for operating and maintenance. The distortion, however, is probably small. CEP assumes that refineries operate on a 365 day year and will operate at an average 90% of capacity over the calculation period. The product of these numbers and estimated refinery capacity yields an estimate for barrels of product per year (line 18). A higher capacity utilization rate raises assumed output and correspondingly lowers fixed cost per product unit. Similarly, a lower assumed use rate raises calculated product costs. Figure A-1 indicates how sensitive the estimate is to assumed use rates.

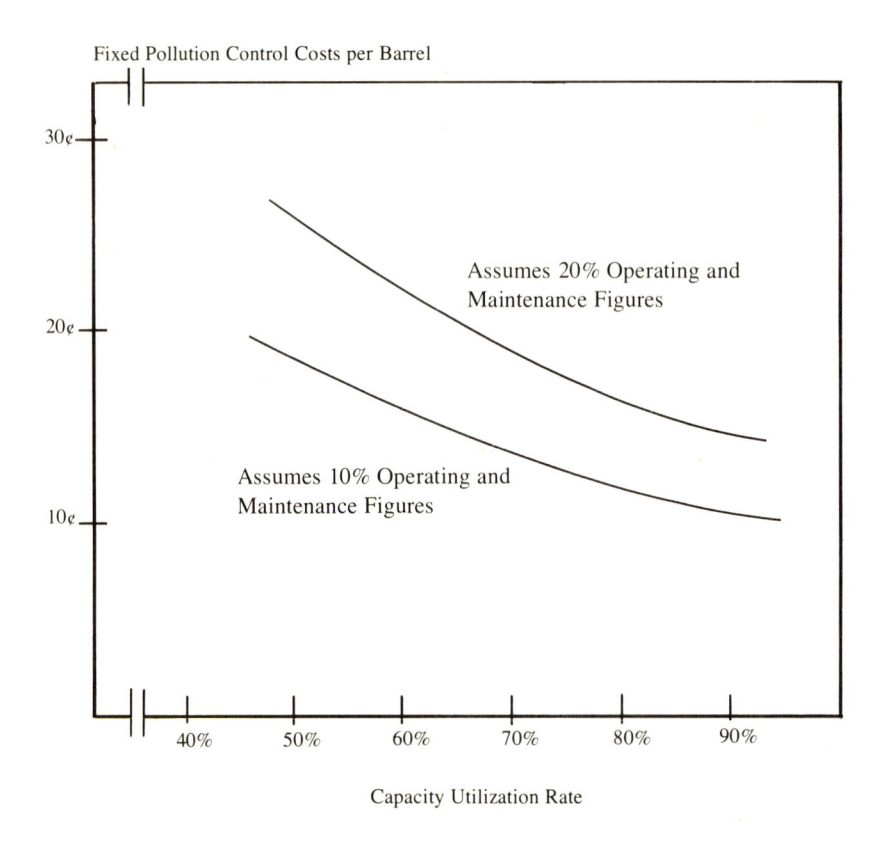

FIGURE A-1. Pollution Control Costs vs. Capacity Utilization Changes.

Total Added Product Costs (lines 19 through 21)

Total added product costs are simply the sum of capital related costs and costs for low sulfur fuels. These latter costs, described in Chapter 4, represent the proxy we have used for increased operating costs. We assume they come to 1.5¢ per barrel of product. The total dollar costs for post-1973 pollution control will vary with output since the fuel estimate is tied to output levels. The unit of product cost result will also vary with assumed output because the capital related costs are fixed for any and all output levels.

Taxes and Other Issues

Added costs must reduce profits or result in higher prices, possibly both. The model described above calculates pre tax costs. These are directly applicable only to the price question. Price increases to recover costs must rise by pre tax cost. This is so because increased revenue from higher prices is subject to income tax.

After tax costs are relevant for profit analysis. Since pollution control costs are, like most business expenses, deductible for income tax purposes, it is the after tax costs that matter. CEP assumes the applicable tax rate is the marginal federal corporate income tax rate, 48%. This ignores state and local income taxes, although they would raise the applicable tax and consequently reduce the effective costs.

The use of 48% may seem strange at first since it is well known that many petroleum companies pay only a small portion of their income in federal income taxes, sometimes as little as 3-4% (23). Such figures are averages, however, and do not represent the rates that the companies pay on marginal, or last dollars of income. Furthermore, they mix in income and expenses from foreign sources with figures from US operations. The rate for petroleum company marginal domestic income is the same as that for all other corporations earning over $25,000 per year, 48%.* Pollution control costs are akin to that last bit of income—they are added on after everything else is done. In this context the applicable tax rate is 48%.

*There is a possibility that pending legislation will change this crucial tax rate. Again, we must ignore this because there are no final results at this time.

SAMPLE CALCULATION

Calculation for the Refining Industry of Added Product Costs due to Pollution Control Installed between 1974 and 1983 at Total Existing Refinery Capacity

(All figures are in millions of 1974 dollars unless otherwise noted.)

Line	Item	1974	1975	1976	1977	1978	1979	1980	1981	1982	1983	1974-1983 TOTAL
1	Required Capital Investment to Upgrade Existing 1974 Capacity (Base capital for calculation.)	340	340	340	340	100	100	100	100	100	100	1960
	Depreciation Calculation (15 year life straight line method.)	23	23 22	23 22 23	23 22 23 23	7 23 22 23 23	7 6 23 22 23 23	7 6 7 23 22 23 23	7 6 7 7 23 22 23 23	7 6 7 7 7 6 23 22 23 23	7 6 7 7 7 6 7 23 22 23 23	
2	Depreciation Total for Year	23	45	68	91	98	104	121	128	134	141	953

Line	Item	1974	1975	1976	1977	1978	1979	1980	1981	1982	1983	1974-1983 TOTAL
		\multicolumn{11}{c	}{Cost of Capital Calculation}	
3	Base Capital Net of ITC	294	294	294	294	86	86	86	87	87	87	8087
4	Gross Capital Invested (Running total of line 3.)	294	588	882	1176	1262	1348	1434	1521	1608	1695	2023
5	Accumulated Depreciation (Sum from previous years + ½ in year.)	12	46	102	182	276	374	490	614	745	882	
6	Net Capital Invested (Line 4 - line 5.)	282	543	780	995	986	974	944	907	863	813	
7	Cost of Capital (25% of line 6.)	71	136	195	249	246	244	236	227	216	203	
		\multicolumn{11}{c	}{Operation and Maintenance Calculations}	
8	Gross Cumulative Capital (Running sum of line 1.)	340	680	1020	1360	1460	1560	1660	1760	1860	1960	13660
9	Operating and Maintenance (Low figure is 10% of line 8.)	34	68	102	136	146	156	166	176	186	196	1366
10	(High figure is 20% of line 8.)	72	136	204	272	292	312	332	352	372	392	2732

Total Pretax Capital Related Cost

11	Depreciation (From line 2.)	23	45	68	91	98	104	121	128	134	141	953
12	Cost of Capital (From line 7.)	71	136	195	249	246	244	236	227	216	203	2023
	Operating and Maintenance											
13	(Low figure is from line 9.)	34	68	102	136	146	156	166	176	186	196	1366
14	(High figure is from line 10.)	72	136	204	272	292	312	332	352	372	392	2732
	Total											
15	(Low figure is lines 11+12+13.)	128	249	365	476	490	504	523	531	536	540	4342
16	(High figure is lines 11+12+14.)	166	317	467	612	636	660	689	707	722	736	5708

Production Calculation

17	Capacity million bbl/d	14.2	14.2	14.2	14.2	14.2	14.2	14.2	14.2	14.2	14.2	
18	Annual Production (Line 16 × 90% × 365.)	4665	4665	4665	4665	4665	4665	4665	4665	4665	4665	46647

Line	Item	1974	1975	1976	1977	1978	1979	1980	1981	1982	1983	1974-1983 TOTAL
				TOTAL ADDED COSTS								
19	Add Low Sulfur Fuel Cost (1.5¢/barrel.)	70	70	70	70	70	70	70	70	70	70	700
	Total											
20	(Low figure is lines 15+19.)	198	319	435	546	560	574	593	601	606	610	5042
21	(High figure is lines 16+19.)	236	387	537	682	706	730	759	777	792	806	6408
	Product Cost, ¢ per barrel											
22	(Low figure is lines 20-18.)	4.2	6.8	9.3	11.7	12.0	12.3	12.7	12.9	13.0	13.1	10.8
23	(High figure is lines 21-18.)	5.1	8.3	11.5	14.6	15.1	15.6	16.3	16.7	17.0	17.3	13.7

Addendum: $1960 million to control 14,200,000 bbl/d is $138 per barrel of daily capacity

Note:
The total numbers in this calculation differ from the totals in the text because the text includes the cost of controlling expansion capacity as well as existing plants.

REFERENCES

1. E. K. Grigsby, E. W. Mills, and D. C. Collins, "Future Capital Requirements for Refined Petroleum Products," paper presented at the National Petroleum Refiners Association Annual Meeting, 1-3 April 1973, San Antonio, Texas.
2. According to Mobil Oil Corporation, "Approximately 10-15 percent [of construction costs] would have to be devoted to pollution control facilities, excluding product desulfurization and lead reduction" (personal communication, 5 June 1974).
3. "The literature reports pollution control costs for refineries which are about 8.3% of the regular cost of refineries. Actual costs and estimates of various authorities range enormously from 1.7% to 17.4% . . . even 20 and 30% upper limits" (W. L. Nelson, "Questions on the Technology of Refinery Pollution Control," *Oil and Gas Journal*, September 4, 1972, page 86.) Nelson's own estimate is given as 7.6%.
4. Shell Oil Corporation, response to CEP economic questionnaire.
5. John E. Cremens, "Capital Expenditures by Business for Air and Water Pollution Abatement, 1973 and Planned 1974," *Survey of Current Business*, July 1974, page 58-64.
6. W. L. Nelson says, for example, ". . . one suspects that high costs were stated to emphasize the cost of ecological facilities" ("Costs of Refineries, Part 4," *Oil and Gas Journal*, July 29, 1974, p. 162).
7. Stephen Sobotka & Company, "Economic Analysis of Proposed Effluent Guidelines, Petroleum Refining Industry, Part I," prepared for the Environmental Protection Agency, Office of Planning and Evaluation, June 1973; and Part II, prepared by the Environmental Protection Agency, September 1973.
8. Environmental Protection Agency, "Development Document for Proposed Effluent Limitations Guidelines and New Source Performance Standards for the Petroleum Refining Industry," December 1973, passim.
9. Ailleen Cantrell, "Annual Refining Survey," *Oil and Gas Journal*, April 1, 1974, pp. 82-105.
10. Robert O. Skamser goes even further and states, "Because of physical and economic obsolescence, 30% of today's refining capacity will be replaced by 1980," ("US Refining Must Double Its Annual Investment for 1980's Needs," *Oil and Gas Journal*, August 27, 1973.
11. Environmental Expenditures of the U.S. Petroleum Industry, 1966-1972." American Petroleum Institute publication #4176, 1973, page 6 et passim.
12. "Petroleum Facts & Figures" (API, 1971), page 470.
13. Brown & Root, Inc., "Economics of Refinery Wastewater Treatment," prepared for the American Petroleum Institute, Committee on Environmental Affairs, API publication #4199, August 1973.
14. Thomas P. Broderick, "An Analysis of Tax Incentives for Certified Pollution Control Facilities," unpublished Masters essay, University of California, Berkeley, 1974, p. 7.
15. Atlantic Richfield Company, Gulf Oil Corporation, Exxon Corporation, Mobil Oil Company, Standard Oil of California, Standard Oil Company (Indiana), Shell Oil Company, and Texaco, Inc., annual reports for 1973.
16. Broderick, page 51-53.
17. Sobotka, 1973, Exhibit 15.
18. Standard Oil Company of California, response to CEP economic questionnaire.
19. Exxon Corporation, response to CEP economic questionnaire.
20. Arco, Exxon, Mobil, Shell, and Standard Oil Company (Indiana) responses to CEP refinery questionnaire.
21. This is the method suggested by the API in "Environmental Expenditures of the US Petroleum Industry, 1966-1972."
22. Albert J. Fritsch and John W. Egan, for example, present these very low figures prominently (*Big Oil: a Citizen's Factbook on the Major Oil Companies*, Center for Science in the Public Interest, 1973).

Glossary

Claus Plant: Used by refineries to convert the hydrogen sulfide generated by refinery processes to pure sulfur, a salable product. It takes its name from the Claus reaction which catalytically burns the hydrogen sulfide to sulfur and water. It is also called a sulfur recovery plant.

Electrostatic Precipitator: A pollution control device that removes particulates from a gas stream by giving them an electrical charge and then collecting them on large metal plates which have an opposing charge.

Finished Fuels Refinery: A refinery whose primary products, such as gasoline and distillate oils, need no further processing before use.

Hydrocarbon: Any of a vast family of compounds made up of hydrogen and carbon. The term is often used to include compounds that also contain elements such as oxygen, nitrogen, or sulfur. Crude oil is a mixture of many kinds of hydrocarbons.

Incremental Costs: The costs of production per unit of product that are added when a new process is used.

Octane Rating: A measure of the antiknock property of a gasoline. Premium gasolines have a higher octane rating than regular gasolines. Many refinery processes, such as catalytic cracking, reforming, and alkylation, are designed to produce high-octane gasoline or to increase the octane rating of gasoline components. The addition of small amounts of additives such as tetraethyl lead also increases the octane rating of most gasolines.

Particulates: Solid or liquid particles in the air: dust, smoke, mist, and spray.

Phenol: A white, crystalline compound found in coal, tar, and crude oil. As a component of wastewater it can be toxic to acquatic life. Its chlorinated compounds produce a foul taste and odor in water.

Photochemical Smog: Formed by reactions between nitrogen oxides and certain hydrocarbons in the presence of sunlight. It is composed of ozone and other photochemical oxidants which irritate the eyes and lungs, damage many plants, and weaken rubber and fabrics.

Sludge: A liquid/solid residue that is formed by some refinery and wastewater treatment processes. Depending upon their composition, sludges are classified as oily, oil-free, chemical, or biological.